# BRITISH CIVIL AVIATION

British Caledonian

# BRITISH CIVIL AVIATION

Günter G. Endres

LONDON

IAN ALLAN LTD

First published 1985

ISBN 0 7110 1552 X



Published by Ian Allan Ltd, Shepperton, Surrey; and printed by Ian Allan Printing Ltd at its works at Coombelands in Runnymede, England

*Cover:*
**The jacket photographs, taken at Heathrow in 1985, were kindly supplied by the British Airports Authority.**

# Contents

**Acknowledgements**

A book of this nature is much more of a joint effort than the sole name of the author on the cover would suggest, and I would herewith like to acknowledge the valuable help of friends, like-minded acquaintances, companies and organisations who have contributed towards this overview of British air transport. If in the last-minute rush to meet deadlines I have inadvertently omitted anyone from this list, I would ask for understanding.

Mike Stroud, John Stroud, John Cook, Mike Hooks, David Dorrell, John Havers, Malcolm Ginsberg, Ron Davies.

Sue Mullen, British Aerospace Hatfield; Harry Holmes, British Aerospace Woodford; Gary Lillystone, British Aerospace Prestwick; British Aerospace Filton; Alice Day, Pilatus Britten-Norman; Kaye Bradford, Westland Helicopters; Bill Purbrick/S. Barbour, Edgley Aircraft; Derek S. Jones/Tom Goyer, Short Brothers; Algy Williams, Airship Industries.

Bruce Hales-Dutton, British Airports Authority; Jerry Scutts, Civil Aviation Authority; Capt E. M. Brown, British Helicopter Advisory Board; Gp Capt L. L. Flowerdew, Air Transport Operators' Association; Capt W. S. Ashpole, British Caledonian Commuter and Airlink; David A. Beety, British Commuter Airlines Association; Mike Kay, Gatwick Regional Airlines' Association; Miss V. R. Harrison, Datapost.

Alan McGregor, Manchester International Airport; Bob Taylor/Philip Roocroft, Birmingham International Airport; James H. Denyer/Christine Moffit, Newcastle Airport; G. C. Dennison, Leeds/Bradford Airport; Ian Cran/Miss P. M. Matthews, Cardiff-Wales Airport; Belfast International Airport; East Midlands Airport; Nicholas Hopkins, John Mowlem & Co.

James L. Croall/Maureen Price, British Airways; British Caledonian Airways; Air UK/Robin Flood Marketing; Paul Reid, Airways International Cymru; Anglo Cargo; J. L. Richardson, British Air Ferries/Guernsey Airlines; Capt R. H. Heath, British Island Airways; British Midland Airways; R. A. Beaumont, Channel Express (Air Services); George Yeaman, Dan-Air Services; Mandy, Euroair; J. M. Budgen, Janus Airways; P. M. E. Rooley, HeavyLift Cargo Airlines; Loganair; Caroline Atkinson, Tradewinds Airways.

Air Eçosse; Trefor G. Jones, Birmingham Executive Airways; Brown Air Services; B. W. Burgess, Connectair; Tony Brunskill, Jersey European Airways; Xenia Woolmer, London European Airways/Raitt Orr & Associates; T. P. R. Liddiard, Manx Airlines; Tony Masefield, Metropolitan Airways; David McCulloch, Spacegrand.

Janet Routledge, Ace Aviation; M. H. Collett, Air Atlantique/General Aviation Services; G. Morrell, Air Service Training; G. S. Mimms, Aviation West/Air Camelot; Mary Brown, CEGA Aviation; Lynn M. Bunton, Colt Executive Aviation; Susan Morley, Directflight; Kay Marshall, Falcon Jet Centre; C. R. Winkworth, Hubbardair; Mike Wilds, IDS Aircraft; J. Keen, Keenair Services; Richard Stone, McAlpine Aviation; Capt I. Slater, Mountleigh Air Services; John G. Gilman, Northair Aviation; Norman Edwards, NEA; Moira J. Fieldwick, Marine & Aviation Management International; Peregrine Air Services; Skycabs; Alan Joyce, Skyguard; Topflight Aviation; Capt M. R. Woodgate, Woodgate Air Services.

Mike Wheatley, Air Hanson; P. N. Hunt, Bristow Helicopters; Douglas Aikman, British Airways Helicopters; C. H. Chambers/P. Strickland, Bond Helicopters; M. T. Etheridge, British Caledonian Helicopters; G. Pryke, Burnthills Highland Helicopter Service; Graham Roberts, DHL International; Valerie Parsons, Dollar Helicopters; Gleneagle Helicopters; R. C. Carpenter, Helicopter Hire; Annette Walls, Skyline Helicopters; Southernair

# British Air Transport – A brief look back

For some years after the Wright brothers' first sustained powered flight in 1903, aircraft were considered of use primarily as military vehicles and for dare-devil joy rides. Once primitive experiments had given way to more controlled technical advancement, progress was swift, soon leading to regular mail and cargo flights, and the odd attempt at carrying passengers. Names still familiar today figure prominently in those early days. Among them J. T. C. Moore-Brabazon, Geoffrey de Havilland, A. V. Roe and C. S. Rolls all took to the air in 1909/10, and in September 1911 Gustav Hamel, flying for the Grahame-White Aviation Co, made the first official mail flight between London and Windsor using a French aircraft. In the midst of war, on 5 October 1916, aircraft manufacturer George Holt Thomas, a man of considerable vision, had founded Britain's first airline, **Aircraft Transport & Travel Ltd**, with the object of 'establishing and maintaining lines . . . for the carriage of mails, passengers, goods and cattle'.

Only days after hostilities had ended, plans were put in hand for a London-Paris service which was inaugurated on 25 August 1919 on a daily basis, using converted de Havilland DH.4 and DH.9A light bombers, which become DH.4As and DH.16s. This red letter day in British air transport also had great significance in a wider context. On the same day a meeting was held (at the instigation of George Holt Thomas and Sefton Brancker) which gave birth to the **International Air Traffic Association (IATA)**. Also represented were the airlines of Germany, Denmark, Norway, Sweden and the Netherlands. Earlier that year, A. V. Roe had started the first domestic air service linking Manchester, Southport and Blackpool, and a Department of Civil Aviation had been formed at the Ministry, with Winston Churchill becoming the first Secretary of State for Air.

Hard on the heels of AT&T came **Handley Page Transport Ltd**, incorporated on 14 June 1919 by Frederick Handley Page, which began regular services from London to Paris and Brussels in September. There were other ventures during the postwar period, but all were short-lived, and even Holt Thomas succumbed to financial difficulties and ceased operations on 17 December 1920. High costs and inadequate revenues forced the other survivors, Handley Page and **S. Instone & Co** (which had begun public services between London and Paris on 18 February 1920), to give up two months later.

In comparison to their heavily subsidised European competitors, British airlines were clearly at a disadvantage, a fact which only slowly dawned on the Government. The sudden cessation of all services rather jolted it out of its lethargy and, with a modest contribution, services were resumed in March 1921 by Handley Page and Instone. In April 1922 the two airlines were joined on the London-Paris route by the **Daimler Airway** which had acquired the assets of AT&T, and all three companies extended their routes to other European cities.

On 2 January 1923 the Government appointed a Civil Air Transport Subsidies Committee under the Chairmanship of Sir Herbert Hambling to consider the working of the cross-Channel subsidies scheme and to advise on long-term solutions. As a result of the Committee's recommendations, a unified national airline was incorporated on 31 March 1924 under the title of **Imperial Airways Ltd**, to develop British air transport on an economic basis. The new airline was formed out of Handley Page Transport Ltd, The Instone Air Line Ltd, the Daimler Airway and **British Marine Air Navigation Co Ltd** (founded on 23 March 1923 to operate cross-Channel amphibian services from Southampton), with a capital of £1million. The new company was to receive a total subsidy of £1million spread over 10 years, and was to use only British aircraft and engines. The United Kingdom thus became the first major country to sponsor its own national airline.

### The Empire Routes

Imperial Airways inherited a good share of traffic on the routes to Paris, Amsterdam and Brussels together with an operational fleet which consisted of three Handley Page W.8bs, seven de Havilland DH.34s, two Supermarine Sea Eagles and one Vickers Vimy Commercial, and a few non-airworthy types. Although other European cities were quickly added to the network, the airline's prime responsibilities were to provide links between Britain and its overseas territories. After preparatory work which

***Top***
**Two Aircraft Transport & Travel aircraft, both photographed in 1919. This DH.16 at Hounslow was the very aircraft which operated the first scheduled 12.30 service to Paris on 25 August 1919, but it was photographed here after a change of registration from K-130 to G-EACT.** *John Stroud*

***Above:***
**DH.4A G-EAJC, which also flew to Paris on 25 August 1919.** *John Stroud*

***Below:***
**An Imperial Airways Armstrong Whitworth Argosy at Croydon Airport.** *British Airways*

included a series of survey flights by Alan Cobham in 1925/26 to India, Burma, Australia and South Africa, Imperial Airways took over the RAF's mail and courier services between Cairo and Baghdad, the so-called 'Desert Air Mail', in January 1927 with an extension to Basra. With departure from London on 30 March 1929, the entire England-India service was opened. This first weekly UK-India service took seven days and was flown with three different aircraft and incorporated a train journey between Basle and Geneva. Imperial Airways reached Singapore on 9 December 1933, and a year later at last fulfilled its dream of a UK-Australia service by opening a regular weekly mail route on 8 December. Total journey time was 12 days. The 12,754-mile route was opened for passengers on 13 April 1935.

Imperial Airways' other major commitment was a route to South Africa. By early 1931 a flying-boat service had pushed the route as far as Lake Victoria, and the Cape Town connection was made on 20 January 1932. Passengers were carried from 27 April that year. The service took 11 days and involved 33 stages and six different aircraft, as well as two train journeys. When the Shorts C-class Empire flying-boats came into service in the late 1930s, the journey was reduced to 4½ days. Services to Britain's West African territories were started in 1936, branching off from Khartoum to Kano via Fort Lamy, extended later that year to Lagos. Services to Hong Kong also began in 1936.

On the North Atlantic, Imperial Airways carried out a number of survey flights and refuelling trials, and opened a mail-only service on 5 August 1939, but the outbreak of war cut short further development in that direction.

Airships also featured briefly in Britain's Empire route plans. In a Government programme two airships were built, the R100 by Vickers and the R101 by the Air Ministry, both of which made their first flight in 1929. The R101 crashed near Beauvais on 5 October 1930 on the first stage of its flight to India, killing 48 people, among them the Air Minister Lord Thomson and Sir Sefton Brancker, Director of Civil Aviation. The crash ended the airship programme.

### Domestic Air Services

After the formation of Imperial Airways in 1924, few domestic air services remained, and it was not until the 1930s that they came back into fashion. Hillman's Airways, formed by Edward Hillman, an omnibus operator, started a service from Romford to Clacton-on-Sea on 1 April 1932 and added a Romford-Paris service in competition with Imperial Airways a year later, using the new de Havilland DH.84 Dragon. Edward Hillman had done much towards the development of the DH.84, from which followed the DH.86 Express and the famous DH.89 Dragon Rapide which formed the backbone of the fleets of most UK airlines in the 1930s and 1940s. Five other airlines were formed in 1932, and many more followed in the next few years, but most were short-lived.

Perhaps the most influential airline of that time was **Railway Air Services**, backed by the four mainline railways and Imperial Airways. RAS began services on 7 May 1934 with a Plymouth-Haldon-Cardiff-Birmingham-Liverpool route, the Plymouth-Birmingham route having been pioneered by the Great Western Railway in 1933, leading to an extensive network of air services throughout southern England with trunk routes through the Midlands to Belfast, Glasgow and the Isle of Man.

On 1 October 1935 Spartan Air Lines and United Airways, both backed by Whitehall Securities, merged to form **British Airways**; and the new airline was joined on 11 December 1935 by Hillman's Airways and on 1 August 1936 by British Continental Airways. From the outset British Airways, now a major carrier, concentrated mainly on the continent of Europe to the detriment of its domestic services. A lack of suitable British aircraft also forced the airline to turn to Fokker, Junkers and Lockheed aircraft for its equipment. In July 1935 a committee was set up to look into the development of civil aviation in the United Kingdom under the Chairmanship of Brig-Gen Sir Henry Maybury. The Maybury Committee reported in December 1936, its principal recommendations being a scheme for licensing all internal airlines with a view to ensuring the most effective service to the public without uneconomical duplication. Another committee under the chairmanship of Lord Cadman criticised Imperial Airways in its report of February 1938 for having neglected its European services and recommended that these should be transferred to British Airways. There was a division of routes and sharing on some, including London-Paris, and this worked reasonably well for a time, but in November 1938 the government of the day decided to amalgamate both airlines into a single State Corporation, **British Overseas Airways Corporation (BOAC)**. The official merger took place on 1 April 1940. Domestic airlines worked together under the umbrella of the **Associated Airways Joint Committee**, formed on 5 May 1940.

Both BOAC and AAJC provided necessary skeleton services during the war years, particularly to Scandinavia, continental Europe, and across the Mediterranean. As soon as conditions permitted, BOAC gradually resumed its former services, initially with Lancastrians (converted

***Above:***
**The famous de Havilland DH.89A Dragon Rapide of Railway Air Services photographed by John Stroud at Croydon and the Bristol Freighter with which Silver City Airways opened the world's first car ferry service in 1948.** *John Stroud*

***Right:***
**British Airways operated several foreign types including this captured Junkers Ju52/3m. Mail is being loaded for the service to Hannover.**
*British Airways*

BRITISH AIRWA
BRITISH AIRWAYS LTD

Lancaster bombers) on the UK-Australia route, and C-class and Sunderland flying-boats, DC-3s, Lockheed 14s and Avro Yorks.

Britain's postwar aircraft development was shaped by the Brabazon Committee, set up by the Government on 23 December 1942 under the chairmanship of Lord Brabazon of Tara. The Committee reported on 9 February 1943 and on subsequent occasions, and recommended the building of nine types of aircraft. From these recommendations evolved such aircraft as the Airspeed Ambassador, Vickers-Armstrongs Viscount, Avro Tudor, Miles Marathon, de Havilland Dove, and the world's first pure jet airliner, the de Havilland Comet 1. A number of others, though magnificent in concept, proved impractical and did not get beyond prototype stage. Among those were the eight-engined, 100-seat Bristol Brabazon, and the 10-engined Saunders-Roe Princess flying-boat.

**Postwar Reconstruction**

A radical restructuring of the British airline industry after the war resulted in the creation of a **British European Airways Division (BEA)** on 1 January 1946. BEA had its beginnings in the activities of No 110 Wing of RAF Transport Command which operated special flights within Europe. These services were taken over on 4 February and operated out of Northolt instead of Croydon. In addition to these and BOAC's European services, BEA also took over, on 1 February 1947, the internal routes of other companies which had resumed independent operations on the disbanding of the Associated Airways Joint Committee.

When BEA became a separate State Corporation on 1 August 1946, Britain had three national airlines, BOAC, BEA and **British South American Airways Corporation**. BSAAC was the successor to British Latin-American Airways Ltd which had been established in January 1944, and operated services to the Caribbean and South America with Avro Lancastrian, York and Tudor aircraft until being absorbed into BOAC on 30 July 1949.

Under the 1939 Air Corporations Act, tightened still further by the new Labour administration in 1945, private airlines were restricted entirely to charter activities. Even so, the availability of relatively cheap war-surplus aircraft, a requirement for military supply operations, and the 1948/49 Berlin civil airlift, gave birth to some 70 companies, of which only a handful survived.

The situation was eased slightly by the Civil Aviation Act of 1949 which permitted the licensing of private companies as associates of the State Corporation, but these were mainly for routes which BEA did not want and were seldom worth having.

The Conservative Government provided further encouragement for the independents by setting up an impartial Air Transport Advisory Council (ATAC) under Lord Terrington, which agreed that the national carriers would seek charter business only when the independents lacked the necessary capacity. Also authorised were special 'Colonial Coach' services to East, West and Central Africa. These low-fare, multi-stop flights, operated by Airwork and Hunting-Clan with Vickers Vikings, were deemed not to make 'material diversion' of passengers from the faster aircraft flown by the state corporations.

Domestic airlines which gained a foothold at this time were Jersey Airlines, Channel Airways, Cambrian Airways, Scottish Airlines, BKS Air Transport, Derby Airways, Eagle Airways, Starways and Dan-Air Services. By 1959 a series of amalgamations had led to the emergence of five major airlines — Airwork, Hunting-Clan, Britavia, Eagle and Skyways — and the field was further narrowed in 1960 when Airwork and Hunting-Clan and their subsidiaries merged to form **British United Airways**. A year later BUA was joined by the Britavia Group which included the two car-ferry airlines, Channel Air Bridge and Silver City Airways, to become the largest privately-owned airline. Eagle meanwhile had been bought by the Cunard Steamship Co and was awarded a route to New York on 21 June 1961. In a reversal of the Air Transport Licensing Board's decision by the Minister, a situation which was to become a regular feature of UK air transport policy in later years, Cunard joined BOAC to operate on the North Atlantic. The enterprise was short-lived and BOAC's monopoly was soon restored. Eagle survived for a few more years but succumbed, as other charter airlines, owing to extreme financial pressures. Eagle, incidentally, had pioneered in the 1950s the 'inclusive-tour' holiday concept, which was to become big business as the years went on.

The two national carriers enjoyed mixed fortunes during the two decades after the war. BOAC had by far the worst of it, highlighted by the Comet 1 crashes in 1954, forcing the hurried introduction of types not really suited to its route structure. BAE on the other hand had more success with the Airspeed Ambassador, and the turboprop Vickers Viscount which brought new standards of comfort and safety to the air and became one of the most successful British airliners ever built. When BOAC eventually caught up by introducing the four-engined Bristol Britannia, dubbed the 'Whispering Giant' in 1957, jet aircraft were already about to make

*Top:*
**The Vickers Viscount helped BEA become a major airline in the 1950s.**

*Above:*
**The Comet 4 was a successful development of the ill-starred Comet 1 airliner. It remained in service with Dan-Air until 1980.** *Dan Air*

*Below:*
**BOAC's 'Whispering Giant', the Bristol Britannia.**
*Bristol*

***Above:***
**Familiar sights in the 1960s were British United Airways (seen here is VC10 G-ATDJ), and British Eagle, which entered into a short-lived association with Cunard on the North Atlantic service.**
*British United Airways*

the long-haul turboprops obsolete. BOAC ordered the Vickers VC10 rear-engined jet airliner in 1957 and, although it became the envy of all and consistently recorded exceptionally high load factors following its introduction on 29 April 1964, it came much too late to make an impact with other airlines. BOAC eventually went over almost entirely to American-built aircraft. BEA, however, continued to operate British aircraft for many years in the shape of Viscounts, Vanguards, Tridents and One-Elevens.

In yet another endeavour to bring some order to Britain's air transport, the Government appointed a committee under Sir Ronald Edwards in 1967 to consider the future organisation of civil aviation. In its report published two years later the Edwards Committee recommended that BEA and BOAC should remain the principal carriers, but that British United Airways should become Britain's 'second force' airline. The 1970 Government White Paper went even further in outlining the complete merger of BOAC and BEA under the title of British Airways, and confirming the BUA option. BUA, operating out of Gatwick, was itself in financial difficulties and was purchased on 30 November 1970 by Caledonian Airways, a non-scheduled operator formed in 1961. The airline, operating as Caledonian/BUA until 1 September 1972 when the name was changed to **British Caledonian Airways**, was awarded the British flag services to West Africa and Libya, and was also licensed to operate between London and Paris.

The Edwards Committee also recommended the creation of a **Civil Aviation Authority** which was to be 'a strong body standing up for British civil aviation . . .'. The CAA assumed its responsibilities on 1 April 1972.

The story of the 1970s would be far from complete without mentioning the world's first supersonic passenger service (which was inaugurated on 26 January 1976 simultaneously by British Airways and Air France between London and Bahrain, and Paris and Washington, with the Anglo/French Concorde), and the name of (Sir) Freddie Laker. Having started in 1948 on the Berlin Airlift, he formed Laker Airways in 1966, and in June 1971 applied to the ATLB to operate a low-fare, no-frills, no-booking scheduled 'Skytrain' service across the North Atlantic. He was of course turned down, and there followed a catalogue of approvals and revocations from both UK and USA authorities which went on until 1977. On 26 September of that year 'Skytrain' finally left Gatwick for New York, bringing regular cheap travel to the prestigious North Atlantic route. Laker's triumph proved by no means the end of his troubles and his airline collapsed on 5 February 1982.

**British Airways**

British Airways came into being on 1 April 1972 when the British Airways Board, set up by the Civil Aviation Act of 1971, assumed control and ownership of the British Overseas Airways Corporation and the British European Airways Corporation. During the next decade British Airways underwent many changes to its operating structure, much of it following the newly-elected Tory Government's proposals in 1979 to privatise the airline as soon as possible, as part of its policy of selling off nationalised companies. However, privatisation had to be delayed when the 1980 oil crisis plunged the airline deeply into the red making it much less attractive to prospective buyers. After a savage cost-cutting programme the airline got back into profit by 1983, and privatisation is now likely in early 1986.

BA's planned sell-off into private hands has been accompanied by much controversy, not least as a result of the Government's rejection of the eagerly awaited CAA review of UK aviation policy, published on 16 July 1984.

Britain has a long history of uncertain policy towards air transport, changing as frequently as Governments and reflecting strongly the different ideologies, usually from an extreme viewpoint. It had been hoped that the CAA review would set a stable pattern for the future by reducing BA's monopoly, increasing competition and giving a boost to private airlines. Powerful political pressures however, led to almost total abandonment of its key points, leaving British Airways virtually unscathed and giving encouragement to British Caledonian. The main points of the White Paper propose a transfer of BA's Saudi Arabian routes to B.Cal, which in turn relinquishes its non-profitable South American routes to BA, together with its dormant licences to Denver and Casablanca, and exclusive rights to Tampa and Orlando.

There is a little crumb of comfort for the smaller independent airlines with some financial help to develop regional services to the Continent. Set against this is the nagging worry about BA's growing influence in the charter market.

***Above:***
**The Laker Skytrain McDonnell Douglas DC-10 and British Airways Concorde were introduced in the 1970s.** *Laker/British Airways*

# British Aircraft – Engineering at its best

Innovation and technical excellence has always been in the forefront of British aircraft manufacture, and air travel throughout the world owes much to the safety and reliability of British aircraft, engines and associated equipment. Most of the 75-year history of British aircraft design and manufacture is embodied in British Aerospace, the company being an amalgamation of most of the famous names of the past, but there are three major companies which have survived under their own names — Short Brothers, the engine manufacturer Rolls-Royce, and Westland, the latter now concentrating entirely on the production of helicopters. Others, like Britten-Norman and Edgley Aircraft, are of more recent origin, but all have added their names to the illustrious list of superior British aircraft.

If this technical superiority has not always been translated into economic success, the underlying reasons are not hard to find. Much of the blame must be laid at the door of successive Governments whose repeated lack of financial support, clear policies and lack of belief in the excellence of Britain's own products has kept the noose firmly tightened around the aircraft industry. Adherence to the strict requirements of the national airlines, too small a market on its own, has also severely limited our export potential. Whilst this has often resulted in a better and more thoroughly tested aircraft, Britain has invariably been beaten to the finishing post by the foreign competition which, more often than not, started much later. There have been exceptions of course, but somehow the timing has never been quite right, and the quantities rarely enough for commercial success. A dwindling Empire has also equated with a shrinking market, though this has been partly redressed, in theory at least, by Britain's membership of the European Economic Community.

These negative aspects, while having to be considered, should not in any way detract from Britain's technological achievements. Many people, even those not normally given to enthusiasm about things that fly, now look up in awe at the sleek, dart-like shape streaking through the skies on its way to the United States, to arrive there less than 3½ hours later. This is Concorde — magnificent in concept, magnificent in execution, magnificent in the air.

**Supersonic Transport**

Concorde, without doubt the world's most advanced airliner, is a joint product of British Aerospace and Aerospatiale of France. It is also the most tested aircraft in the history of civil aviation, being the only SST in the world to receive a Certification of Airworthiness, to allow passenger flights, from the internationally-recognised authorities of Britain, France and the United States.

The world's first supersonic passenger services began on 21 January 1976, when British Airways and Air France Concordes flew to Bahrain and Rio de Janeiro respectively. Concorde services were later also provided to Caracas, Mexico City, Dallas/Fort Worth, Washington, New York and Singapore, but some of these have since been discontinued. Currently, British Airways operates a twice-daily Concorde service to New York and three direct flights a week to Washington, which then continue on to Miami. Air France runs seven supersonic services a week to New York. Many passengers are also carried on charter flights, for which there is a growing demand.

Concorde's history goes back to the late 1950s when independent design studies into supersonic transport were initiated by Bristol Aircraft in the UK and Sud-Aviation (later to become Aerospatiale) in France. Financial considerations led to a pooling of resources, and on 29 January 1962 an agreement was reached between the British Aircraft Corporation (successor to Bristol), Rolls-Royce, Sud-Aviation and SNECMA (the French engine manufacturer) to jointly develop a supersonic airliner, appropriately named 'Concorde'. This decision also marked the first major collaborative venture in Europe from which many more were to follow.

All design, development and construction work was shared between these participants, and the end product was a 120-seat delta-wing aircraft powered by four specially-designed Rolls-Royce/SNECMA Olympus turbojets, and capable of flying at more than twice the speed of sound, thus halving flight times on key international routes. Two prototypes were constructed, 001 (F-WTSS) at Toulouse and 002 (G-BSST) at Bristol. The two aircraft made their maiden flights on 2 March and 9 April 1969 respectively. Elsewhere in the world, Boeing was working on a much larger and faster SST, the

***Top:***
**One of two prototypes of Concorde carrying the original names of the two manufacturers which were to become Aerospatiale and British Aerospace.** *British Airways*

***Above:***
**There is no greater sight in civil aviation than Concorde . . . except perhaps two Concordes.**

***Above:***
**The Kingston headquarters of the British Aerospace Aircraft Group.** *BAe via Mike Stroud*

***Below:***
**The origins of British Aerospace.**

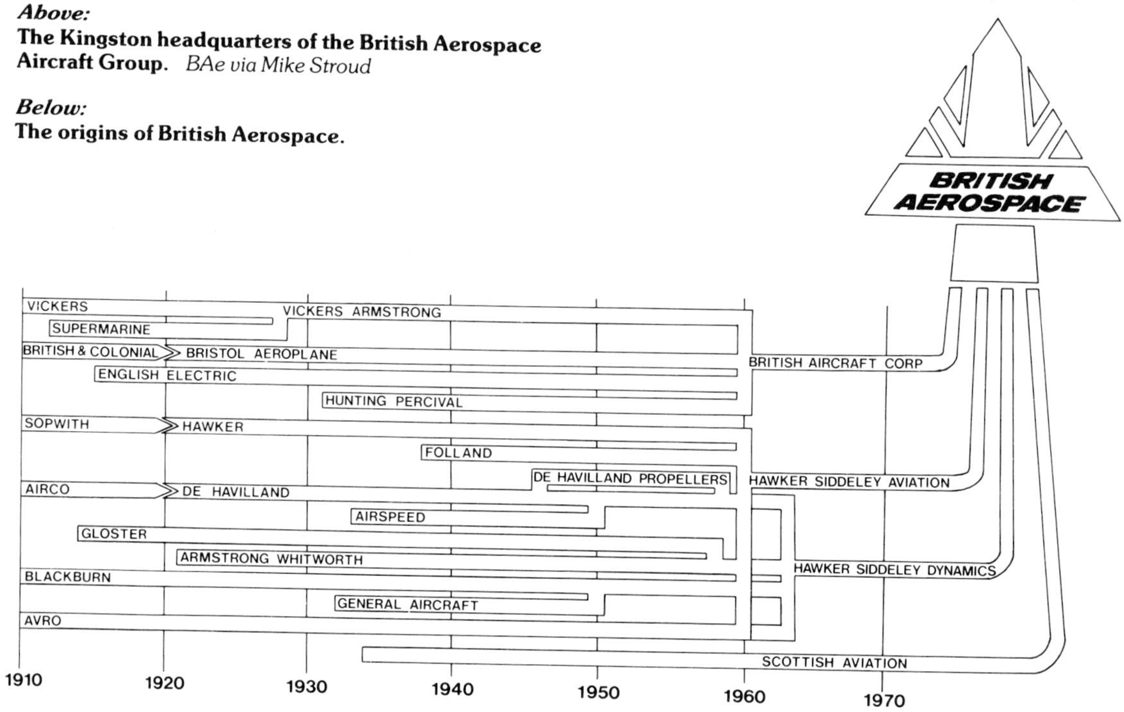

2707-300, but all development was halted in 1971. In the Soviet Union the Tupolev Tu-144, much closer in size and performance, flew first on 31 December 1968, beating Concorde into the air by two months, but did not enter limited passenger services with Aeroflot to Russia's Far East until almost two years after Concorde. Aeroflot encountered major technical problems with the Tu-144, forcing a complete withdrawal from service by around 1978.

The arrival of Concorde generated strong feelings, both for and against. Its supporters felt a surge of pride for their countries' achievement, while the environmentalists, worried about the noise and effect on the atmosphere, united against the project. Commercial prospects nevertheless looked good, with 74 aircraft ordered or optioned by some of the world's major airlines.

By the time Air France and British Airways inaugurated scheduled services, rapidly escalating costs of aircraft and fuel, and the anti-Concorde lobby, had reduced the order book to the aircraft ordered by these two airlines. That, alas, is how it was to remain, and production ceased in 1979 after 16 aircraft. Concorde still has strong opposition, but the fact that the aircraft is a supreme technical achievement is beyond question. Nine years of accident-free operation, 80,000 flying hours, 125 million kilometres covered and 1.25 million passengers transported are ample proof of that.

## THE SOCIETY OF BRITISH AEROSPACE COMPANIES

The SBAC is the trade association of Britain's aerospace industry. It was established in 1916 under the title of The Society of British Aircraft Constructors when aircraft development and production was in its infancy. Today, with an international reputation as the trade association of the second largest aerospace industry in the Western world, it can proudly claim close involvement with every major development in Britain's aerospace industry over nearly three-quarters of a century.

While the abbreviated letters SBAC, by which the society is generally known, have been retained, the name was changed in 1964 to The Society of British Aerospace Companies, to reflect more fully the growing involvement of its companies in the manufacture of vehicles and systems for space applications.

Membership of the society currently comprises some 300 companies representing the industry's wide range of activities from airframes, aero-engines, guided weapons and satellites to engineering and electronic equipment. The aim of the SBAC is to serve the best interests of its member companies towards expanding the trade and prosperity of the industry. The society seeks to present and maintain initiatives with the government of the United Kingdom, the industry's major customer, speak with a strong voice in the military and commercial aviation world of Europe and develop and maintain markets all over the world.

The SBAC organises the world-renowned Farnborough International Exhibition and Flying Display, now established as a major event in the international aviation calendar.

The Technical Department of the SBAC co-ordinates work on standardisation, research and development, airworthiness and quality control within the industry. Policy is evolved through various committees, whose members are drawn from the industry, and who advise on a wide range of subjects, including technical matters, sales and exports, equipment, materials, and the specialist fields of contracts, taxation, liaison with Europe, flying regulations, production and education and training.

## BRITISH AEROSPACE

British Aerospace is now one of the world's largest and most powerful aerospace enterprises, employing some 76,000 people in the UK and overseas, and offering a range of products unequalled by any other company. In 1983 BAe's sales rose to £2,300million, and the forward order book to £4,907million. Exports accounted for 62% of total sales.

The structure of BAe was changed in February 1981 from public ownership to a public limited company in the private sector, and the resultant ownership of British Aerospace is now divided between the Government with 48.43%, private shareholders 48.43% and employees with 3.14%. On 15 January 1985 the government announced that it would place most of its remaining shares on the open market.

Although the company is less than 10 years old, having been established on 29 April 1977 through the merger of Hawker Siddeley Aviation and the British Aircraft Corporation, its pedigree embraces almost all the illustrious firms which pioneered and developed aircraft in this country, and put the great into Britain's achievements in aviation technology. A few famous names remained separate, and others such as Handley Page and Miles, alas, fell by the wayside.

It was perhaps appropriate that British Aerospace would establish its headquarters at Brooklands, Weybridge, Surrey, where it all began in the summer of 1908. For it was at Brooklands that Alliott Verdon-Roe, a Manchester man, became the first Englishman to fly a powered aircraft. The fence on which he tethered his flimsy machine remains (as does the AVRO name, later to become part of Hawker Siddeley Aviation); marking the spot still stands a plaque erected by Vickers-Armstrong the inheritors of the site, which became the centre for the British Aircraft Corporation.

Following on from these early tentative steps came a succession of pioneers including some of the best known names in British aviation. Among them were T. O. M. (later Sir Thomas) Sopwith, Robert Blackburn, H. P. Folland, Geoffrey (later Sir Geoffrey) de Havilland, Harry G. Hawker, John (later Sir John) Siddeley and George Holt Thomas, most of whom had formed their own companies by the outbreak of World War 1. Loss of military orders at the end of hostilities put many companies into difficulties, and the 1920s and 1930s witnessed a certain measure of consolidation, but also the foundation of new companies such as Airspeed, Gloster, General Aircraft, Percival and Folland.

By far the largest output of aircraft in the interwar years was for military purposes, but a range of civil airliners was nevertheless produced, mainly for Imperial Airways. The most successful in this field was de Havilland with many noteworthy types, such as the DH.34, DH.50, DH.66 Hercules, DH.84 Dragon, DH.86 Express and the DH.89 Dragon Rapide, all biplanes, a great number of which were also exported. Armstrong Whitworth concentrated on large aircraft such as the 20-passenger Argosy, 40-passenger Ensign and the 17-seat Atalanta, the latter serving on African and Indian routes until 1939. Others of note were the Vickers Vimy Commercial, Supermarine Sea Eagle and the Bristol Ten-Seater.

The end of World War 2 marked the beginning of a significant technological revolution — the advent of the gas turbine engine. In the civil field Britain led the way with the world's first turboprop airliner to enter production, the Vickers-Armstrong Viscount, and the first jet airliner to enter service, the de Havilland Comet 1. A developed version, the Comet 4, opened the first regular jet services between London and New York on 4 October 1958, and remained in service for over 20 years.

It had by then become apparent that a measure of rationalisation was necessary in order to ensure the survival and future stability of Britain's aircraft industry. Thus, in accordance with government policy, Hawker Siddeley was further strengthened with the acquisition of the Blackburn, Folland and de Havilland companies in 1960, and Bristol, English Electric, Vickers-Armstrong and Hunting-Percival came together to form the British Aircraft Corporation. The intention to further consolidate the industry by merging Hawker Siddeley and BAC was first discussed as early as 1967, but 10 years were to elapse before this became a reality. A Bill entitled 'The Aircraft & Shipbuilding Industry Bill' had a rough passage through Parliament but was finally approved in 1977. On 29 April 1977, the merged companies became the British Aerospace Corporation.

## BAe Aircraft Group

British Aerospace is split into two major operating groups, the Aircraft Group and Dynamics Group. The Aircraft Group's main programmes include the design, development, production, marketing, overhaul and support of civil and military aircraft, and the provision of defence support services, such as maintenance and training to overseas countries. Civil aircraft comprise a complete family of airliners and business aircraft, and military programmes range from basic trainers to advanced strike aircraft, as well as specialised maritime patrol and early airborne warning aircraft.

The Dynamics Group is one of Western Europe's major defence contractors, and is also a major force in European space engineering with particular expertise in communications satellites, scientific satellites, solar power generation and spacecraft sub-systems.

The Aircraft Group, employing some 58,000 people, is headquartered at Kingston-upon-Thames and organised into three divisions (Civil, Warton and Weybridge), operating from 14 locations in the United Kingdom. Final assembly of civil aircraft is undertaken at Hatfield, Prestwick, Chester and Woodford, and currently includes the BAe 146 feederjet, the A320 Airbus narrow-body airliner, the Jetstream 31 light turboprop transport, the 125-800 business jet, and the new ATP-Advanced Turboprop. These augment the continuing range of the successful A300/A310 Airbus family, the BAC One-Eleven twin-jet, and the 748 turboprop transport.

British Aerospace is also a leader in international collaboration and presently has partnership agreements with companies in 19 countries on a wide variety of co-operative programmes. These include a 20% membership of Airbus Industrie, with BAe designing and building the wings; the Tornado combat aircraft built jointly with Germany and Italy; Jaguar with France and India; Hawk with Finland and the USA; Harrier with the USA; 748 with India; and the One-Eleven with Romania.

## BAe 146

Flagship of the current civil range is the BAe 146 short-haul regional airliner. A high-wing cantilever monoplane, the 146 is powered by four

*Below:*
**Final assembly line of the BAe 146 at Hatfield.** *BAe*

*Below:*
**Dan-Air is the only UK operator of the BAe 146 feeder jet, with four Series 100 in service on domestic and European routes.** *BAe*

Avco Lycoming ALF 502 high by-pass ratio turbofans which contribute to the aircraft's low noise levels, high fuel efficiency and economy of operation. Wide-body comfort for up to 111 passengers and outstanding airfield performance are bringing new standards of travel to those routes previously operated mostly by propeller-driven airliners and first generation narrow-bodied jets.

The BAe 146 is presently being built in two versions. The Series 100, seating 82-93 passengers, can carry its maximum load from sea-level runways of 1,200m in length and offers exceptional performance from high-altitude, high-temperature airfields. The aircraft can operate from both unpaved and low bearing-strength runways with minimal ground facilities. The Series 200, with a longer fuselage and increased weights, seats 100-111 passengers. A new stretched 120-seat derivative, the 146-300, is due to fly in 1987 with deliveries to customers planned for early 1988. There are also plans for mixed passenger/freight, all freight and military versions. Design and final assembly of the 146 is undertaken at Hatfield with major sub-assemblies being produced at four other BAe factories, and by Avco Aerostructures at Nashville, Tennessee, by SAAB-Scania at Linköping, Sweden, and by Shorts at Belfast.

Hawker Siddeley Aviation's proposals for a quiet 70-100 seat aircraft for short-haul feeder routes, known as the HS.146, received government support on 29 August 1973, but a deterioration in the UK economy fuelled by the oil crisis halted the programme only a few months later. Development work however continued on a limited basis over the next few years, and full go-ahead for the 146 programme was given by the British Aerospace Board with government approval on 10 July 1978.

Roll-out of the Series 100 prototype (G-SSSH) took place on 20 May 1981 and the first flight was made on 3 September 1981. This aircraft together with G-SSHH (first flight 25 January 1982) and G-SSCH (first flight 2 April 1982) amassed a total of 1,500 hours of testing between the first flight and the award of full CAA certification on 4 February 1983. A fourth aircraft (G-SCHH) was used on route-proving. The 146 was the first aircraft in the world to be certificated to the European Joint Airworthiness Requirements (JAR). The BAe 146-100 entered service with Dan-Air, the British independent airline, on 27 May 1983 and is currently used on scheduled services from London to Berne, Toulouse, Dublin, Newcastle, Stavanger and Bergen, and from Newcastle to Amsterdam, Bristol, Cardiff and the Channel Islands. The Series 200 was put into service by Air Wisconsin on 27 June 1983 operating a high-frequency schedule linking Toledo, Fort Wayne and Appleton to Chicago O'Hare International.

The largest single and most significant order to date for the 146, valued at £200million, was placed by the US West Coast operator, Pacific Southwest Airlines (PSA) in November 1983.

**Orders and options for the BAe 146 (as at 1 July 1985)**

| *Operator* | *Series* | *Ordered* | *Options* | *Delivered* |
|---|---|---|---|---|
| AirPac Airlines, Seattle, Washington, USA | -100 | 1 | 1 | 1 |
| Air Wisconsin, Appleton, Wisconsin, USA | -200 | 7 | 6 | 6 |
| Airlines of Western Australia, Perth, Australia | -200 | 2 | 6 | 2 |
| Dan-Air Services, London, UK | -100 | 3 | 2 | 4* |
| Govt of Mali, Bamako, Mali | -100 | 1 | – | 1 |
| Pacific Southwest Airlines, San Diego, California, USA | -200 | 20 | 25 | 10 |
| Royal Air Force, UK | -100 | 4 | – | 2 |
| TABA, Manaus, Brazil | -100 | 2 | 3 | 2 |
| Aspen Airways, Denver, Colorado, USA | -100 | 2 | – | 2 |
| CAAC, Beijing, China | -100 | 10 | – | – |
| | | 52 | 43 | 30 |

*includes one aircraft leased from BAe

The airline chose the 200 version with 100 seats and ordered 20 aircraft for delivery in 1984 and 1985 with options for 25 more.

**BAe Jetstream 31**

British Aerospace announced the launching of the Jetstream 31 light turboprop transport on 5 December 1978, and the development aircraft (G-JSSD), converted from an earlier Jetstream 1 built by Handley Page, made its maiden flight on 28 March 1980.

The decision to go ahead with full series production was made in January 1981 and this is taking place at the BAe factory at Prestwick. Ninety aircraft have so far been authorised for production, over 50 of which have now flown. The production rate for the Jetstream 31 is being built up to 36 aircraft a year by 1986.

The first production aircraft (G-TALL) was rolled out at Prestwick on 25 January 1982 and made its first flight on 18 March 1982. The UK Certificate of Airworthiness was granted on 29 June 1982 and was followed by the United States FAA Type Certificate on 30 November 1982. Initial deliveries to German and UK customers commenced in December 1982.

The Jetstream 31, powered by two Garrett TPE 331-10 turboprop engines, is available in three prime versions:

Jetstream Commuter is the basic model to accommodate 18/19 passengers and baggage. It can carry 18 passengers over ranges up to 1,250km with full IFR reserves.

Jetstream Corporate is designed for operation as an 8/10 passenger executive aircraft capable of carrying nine passengers in luxury up to 1,970km.

Jetstream Executive Shuttle is intended primarily for the business charter market. A wide choice of interior arrangements is possible giving a range of up to 1,880km with a typical layout for 12 executives.

A Quick Change (QC) fit is also available enabling the 8/9-seat version to be converted into a 12-seat executive shuttle or 18-seat airliner layout in only a few hours. The aircraft is also adaptable for various military uses and the Royal Navy has ordered four Jetstream 31s as observer trainers. It can also be used in other quasi-military roles such as communictions, radar operator training, airfield calibration, freighting, casualty evacuation and flight check work. A new development, designated **Jetstream 31EZ (Economic Zone)**, was announced in September 1982. This version has been proposed to meet the growing worldwide need for a cost-effective aircraft capable of patrolling the 200-mile (320km) exclusive economic Zones declared by many countries around their shores.

Thirty-eight aircraft are currently in service worldwide with commuter airlines, executive and charter operators and the military. In the United Kingdom, the Jetstream is used by London European Airways, Birmingham Executive Airways, Peregrine Air Services, McAlpine Aviation and the Royal Navy.

**Orders for the Jetstream 31 (as at 1 July 1985)**

| *Operator* | *Ordered* |
|---|---|
| Contactair, Stuttgart, West Germany | 3 |
| Peregrine Air Services, Aberdeen, UK | 2 |
| Birmingham Executive Airways, Birmingham, UK | 3 |
| McAlpine Aviation, Luton, UK | 5* |
| Atlantis Airlines, Florence, South Carolina, USA | 8 |
| Skywest Airlines, Perth, Western Australia | 5 |
| UK Corporate Operator | 2 |
| US Corporate Operator | 2 |
| Jetstream International Airlines, Pennsylvania, USA | 6 |
| Eastern Metro Express, Atlanta, Georgia, USA | 12 |
| Royal Navy | 4 |
| Netherlines, Rotterdam, Netherlands | 4 |
| Swedair, Stockholm, Sweden | 1 |
| American Eagle, Houston, Texas, USA | 10 |
| Republic Express, Memphis, Tennessee, USA | 20 |
| | 87 |

In addition, BAe Civil Division at Prestwick operates its own demonstrator.

* McAlpine Aviation has been appointed a corporate completion centre for Jetstream 31. Four of the five Jetstream 31s have been completed to date.

*Top:*
**The Jetstream production line at Prestwick, Scotland.** *BAe*

*Above:*
**Peregrine Air Services was the first UK operator of the Jetstream 31. The company flies executive and oil-related charters from Aberdeen Airport.** *BAe*

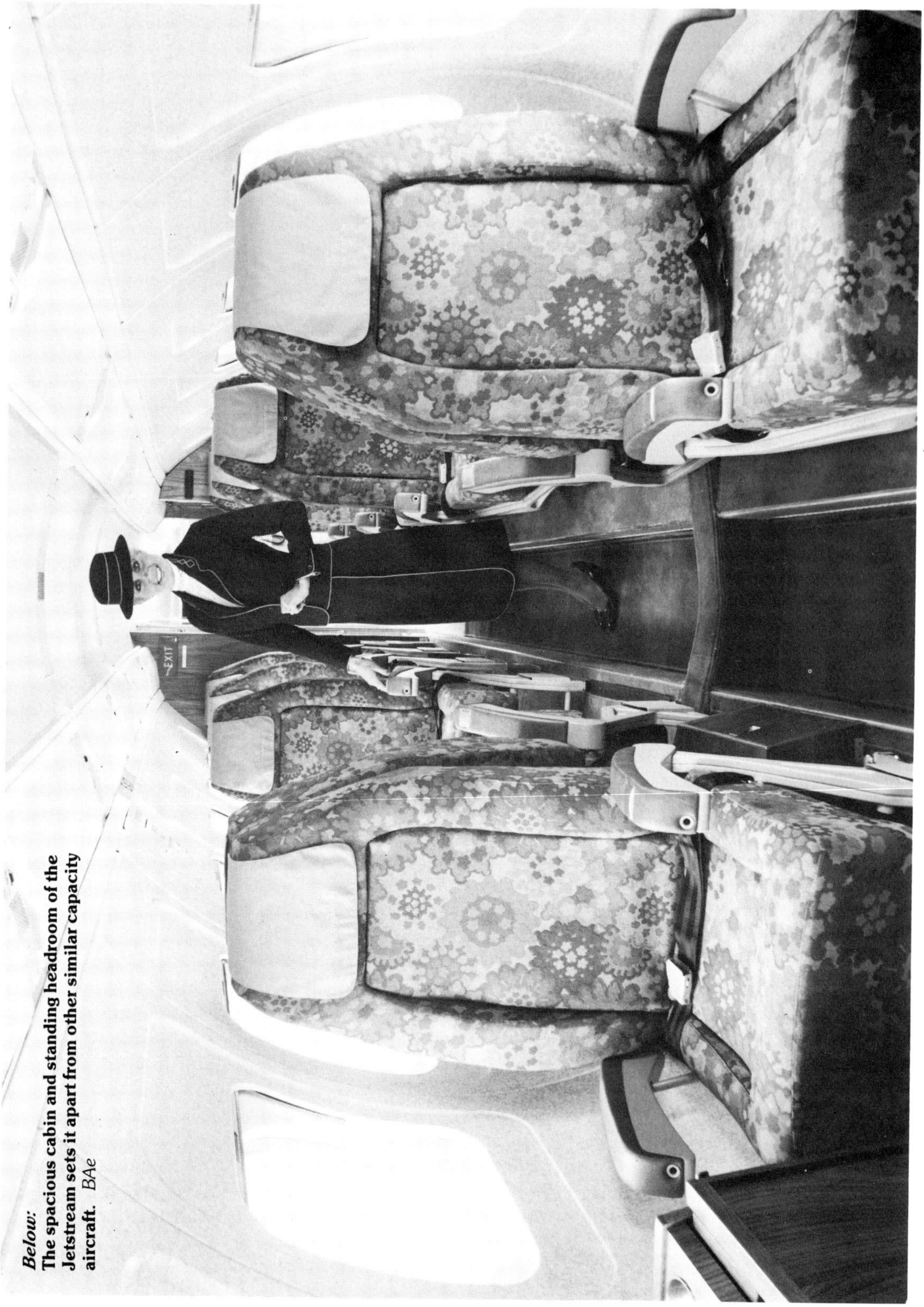

*Below:*
**The spacious cabin and standing headroom of the Jetstream sets it apart from other similar capacity aircraft.** *BAe*

**BAe 125**

Built at the company's Chester factory, the 125 business jet was originally designed by a de Havilland design team under the designation DH.125. The prototype (G-ARYA) first flew on 13 August 1962 powered by two Rolls-Royce Viper 20 turbojets. The introduction of higher-thrust Viper 521 and 522 engines soon resulted in the Series 1A/1B aircraft. The Series 2 was produced for the Royal Air Force as a navigation trainer, and the later Series 3 and 400 differed primarily in having more powerful engines. A lengthened fuselage to accommodate up to 14 passengers accounted for the model 600, and the next major change came with the Series 700 in 1976 which introduced the fuel-efficient Garrett turbofan engines and many other detailed improvements. Since the first delivery was made in September 1964, over 600 BAe 125s have been sold in 33 countries around the world, including some 350 in the highly competitive North American market. Approximately 80% of sales have been for export.

The latest version is the Series 800, the prototype of which (G-BKTF) made its first flight at BAe's Chester factory on 26 May 1983. It is the most radically improved version to date, giving considerably enhanced performance, particularly in terms of range and speed, enabling it to fly non-stop coast-to-coast across the United States, and also giving it trans-Atlantic capability. Certification of the Series 800 was obtained in May 1984.

It is as a corporate jet transport for up to 14 passengers that the 125 has made its mark, especially in North America, but its capability as a high speed light transport makes it also particularly attractive for military and government communication purposes. Other roles to which the aircraft's versatility is especially suited include air ambulance, airways calibration, maritime surveillance, and photo and geophysical survey.

**BAe 748**

The British Aerospace 748 is a versatile, robust and reliable twin-turboprop transport which can carry 44-52 passengers in pressurised and air-conditioned comfort, or six tonnes of freight. Manufactured by the Manchester factories of the Weybridge Division, this well-proven aircraft, powered by two Rolls-Royce Dart engines, has consistently achieved worldwide sales success with 373 sold to 79 different operators in 50 countries. Over 80% of production has been for export, and sales of the 748 have earned more than £400million in foreign currency.

The 748 started life as an Avro project in 1958, and the first prototype (G-APZV) was flown at Woodford, Cheshire on 24 June 1960. The first production Series 1 aircraft with accommodation for 48 passengers and Dart 514 engines was certificated in December 1961, and was followed by the Series 2 and 2A with more powerful engines. These two versions account for the great majority of aircraft built. In 1979 the 748-2A was replaced with the Series 2B with improved 'hot and high' Dart 536-2 engines and various aerodynamic refinements.

The latest version, the Super 748 first announced in February 1982, is now available to customers. This aircraft incorporates a number of significant new developments over the existing 748-2B model, including a new advanced flight deck, new-style galley and installation of Dart engine hush kits and Standby Water Methanol System. Hush kits can be retrofitted to earlier Series 2A and 2B aircraft.

All 748s are available in civil and military transport and Coastguarder versions. The military version has a large rear freight door which can also be fitted to the civil transport, and the Coastguarder is a variant for search and rescue and maritime surveillance duties. To date 748 operators have logged over four million flying hours and have completed 4.5 million landings.

The 748 is subject to a manufacturing agreement with the Indian government, and 89 aircraft have been assembled by Hindustan Aeronautics at Kanpur from British-built components. Of these, 17 were for Indian Airlines and 72 for the Indian Air Force.

**British Aerospace ATP**

The Advanced Turboprop aircraft (ATP) is a second generation twin-turboprop airliner that takes full advantage of advances in technology to bring to the regional airline market new standards in low operating costs, low noise levels, passenger comfort and high profit potential. The ATP, with a seating capacity of 64, has been designed to fill the gap between the highly-successful 44-52 seat 748, to which it clearly owes its origins, and the 82-120 seat BAe 146 family of jetliners.

Work on the aircraft started in 1980 following in-depth research by BAe Manchester into all turboprop routes, which highlighted a requirement for a 60-plus seat, low-noise turboprop offering lowest possible operating costs. The strongest demand for such an aircraft was shown to be in Europe and North America, with other good possibilities in Australasia. BAe believes that over the next decade a total of 500-plus aircraft will be required, of which it hopes for a market share of between 30% and 50%, depending on the number of competing aircraft.

*Top:*
**Production of the 125 business jet is undertaken at the British Aerospace factory at Chester. Approximately three aircraft are produced each month. The photo shows a line-up of 125-700s.** *BAe*

*Above:*
**Current production version of the 125 is the improved Series 800. G-BKUW made the first non-stop flight across the North Atlantic.** *BAe*

*Above:*
**British Airways' first 748-2B Super, in the new livery, outside BAe Manchester prior to delivery.** *BAe*

***Left:***
**BAe's successful 748 turboprop transport is still being produced 25 years after its first flight.** *BAe*

***Above:***
**An artist's impression of the ATP-Advanced Passenger Transport, due to enter service in 1987, and the fuselage of the first aircraft under construction.** *BAe*

Launch customer for the ATP was LIAT of Antigua, which has ordered two aircraft and this was followed with an order for five by the British Midland group.

The ATP will be powered by two Pratt & Whitney PW124 engines each of 2,520 TEHP. These new fuel-efficient engines are presently being flight tested on board a Viscount flying test-bed, and are due to be delivered in October 1985 for marrying to the fuselage, currently under construction. The first flight of the ATP has been set for 6 August 1986, with an in-service date of September 1987. Certification will be to the latest European and United States airworthiness standards.

**BAC One-Eleven**

The One-Eleven twin-jet airliner family is a product of the Weybridge Division, the most recent versions being the 99-119 seat One-Eleven 500 and the 74-89 seat high-performance, short-field One-Eleven 475.

Originally launched in 1961, the prototype One-Eleven 200 (G-ASHG) flew for the first time at Hurn Airport on 20 August 1963, and the type entered service with British United Airways on 9 April 1965. Steady developments offering enhanced capabilities led to the Series 300 with more powerful engines and greater fuel capacity, and the Series 400 incorporating modifications to meet US requirements. The enlarged Series 500 with accommodation for up to 119 passengers first appeared in June 1967, and the final variant, the Series 475, in August 1970. The last-named retains the standard fuselage of the 400, but has the powerplant and wings of the Series 500. All versions are powered by two Rolls-Royce Spey turbofans. Notable in-service developments include a freight door, a Category IIIA automatic landing system, a new-look wide-body interior, and an extensive range of conversions. Total One-Eleven sales now stand at 230 aircraft. Apart from serving with major airlines, the One-Eleven is also used by many business corporations, providing top-class service in many parts of the world, notably in the USA, Europe, the Middle East, Far East and Australia. In the largest technology transfer agreement in the history of civil aircraft, concluded in June 1979, Romania has embarked upon the building of the One-Eleven under licence in a deal worth in excess of £150million. Under the ROMBAC plan three complete aircraft — two Series 500 and one Series 475 freighter — were built in British Aerospace's own factories to initiate the transfer education process, with the first aircraft being completed on the Romanian assembly line in September 1982. A reducing scale of UK supplied kits of parts over the first 22 aircraft in seven stages will complete the transfer by 1987. Thereafter, manufacturing of the complete aircraft in Romania will continue as the market demands into the 1990s.

**Airbus**

Although Britain did not become a full member of the Airbus consortium until 1978, it has been involved since the 1960s, when British manufacturers were invited to participate in a subcontract role with particular responsibility for the design and manufacture of the wings and with assistance in marketing and support. The British share represented by British Aerospace is 20%, the French (Aerospatiale) 37.9%, the German (MBB) 37.9% and the Spanish (CASA) 4.2%. Although BAe's 20% interest was purchased with assistance from the government, the company has also invested substantial sums in the programme, totalling over £250million, from its own resources.

The first A300 (F-WUAB) made its maiden flight from Toulouse on 28 October 1972, and

| | **BAe 146-200** | **Jetstream 31** | **BAe 125-800** |
|---|---|---|---|
| *Dimensions* | | | |
| Wing span | 26.34m | 15.85m | 15.66m |
| Length | 17.81m | 14.37m | 15.60m |
| Height | 8.61m | 5.32m | 5.36m |
| Wing area | 77.30sq m | 25.08sq m | 34.75sq m |
| *Weights* | | | |
| Maximum take-off | 40,597kg | 6,600kg | 12,430kg |
| Maximum payload | 10,206kg | 1,560kg | 1,088kg |
| *Powerplant* | 4×31.0kN Avco Lycoming ALF 502R-5 turbofans | 2×671kW Garrett AiResearch TPE331-10 turboprops | 2×19.13kN Garrett AiResearch TFE731-5R-1H turbofans |

| | | | |
|---|---|---|---|
| *Performance* | | | |
| Maximum cruising speed | 778km/hr | 488km/hr | 858km/hr |
| Take-off field length | 1,612m | 1,326m | 1,645m |
| Range with max payload | 1,853km | 1,038km | 5,486km |
| Range with maximum fuel | 2,743km | 1,167km | 5,680km |
| Service ceiling | | 9,630m | 13,100m |

| | **BAe 748-2B Super** | **Rombac/BAC One-Eleven 475** | **Rombac/BAC One-Eleven 560** |
|---|---|---|---|
| *Dimensions* | | | |
| Wing span | 31.23m | 28.50m | 28.50m |
| Length | 20.42m | 28.50m | 32.61m |
| Height | 7.57m | 7.47m | 7.47m |
| Wing area | 77.00sq m | 95.78sq m | 95.78sq m |
| *Weights* | | | |
| Maximum take-off | 21,092kg | 44,680kg | 47,400kg |
| Maximum payload | 5,257kg | 10,733kg | 11,474kg |
| *Powerplant* | 2×1,700kW Rolls-Royce Dart 536-2 turboprops | 2×55.8kN Rolls-Royce Spey Mk 512-14DW turbofans | 2×55.8kN Rolls-Royce Spey Mk 512-14DW turbofans |
| *Performance* | | | |
| Maximum cruising speed | 452km/hr | 871km/hr | 871km/hr |
| Take-off field length | 1,393m | 1,676m | 1,981m |
| Range with max payload | 1,455km | 2,694km | 2,459km |
| Range with maximum fuel | 2,372km | 3,582km | 3,515km |
| Service ceiling | 7,620m | 10,670m | 10,670m |

| | **Concorde** | **Airbus A300B4-600** | **Airbus A310** |
|---|---|---|---|
| *Dimensions* | | | |
| Wing span | 25.55m | 44.84m | 43.90m |
| Length | 62.10m | 53.62m | 47.21m |
| Height | 11.40m | 16.53m | 15.81m |
| Wing area | 358.22sq m | 260.00sq m | 219.89sq m |
| *Weights* | | | |
| Maximum take-off | 185,066kg | 165,000kg | 132,000kg |
| Maximum payload | 11,340kg | 42,900kg | 34,700kg |
| *Powerplant* | 4×169kN RR/SNECMA Olympus 593 Mk 610 turbojets | 2×249kN P&W JT9D-7R4H1 or 2×249kN GE CF6-80C2 turbofans | 2×214kN P&W JT9D-7R4D1 or 2×223kN GE CF6-80A3 turbofans |
| *Performance* | | | |
| Maximum cruising speed | 2,179km/hr | 891km/hr | 897km/hr |
| Take-off field length | 3,400m | 2,560m | 1,783m |
| Range with max payload | 6,228km | 5,340km | 8,170km |
| Range with maximum fuel | 6,582km | 5,930km | 8,670km |
| Service ceiling | 18,290m | 10,675m | 10,675m |

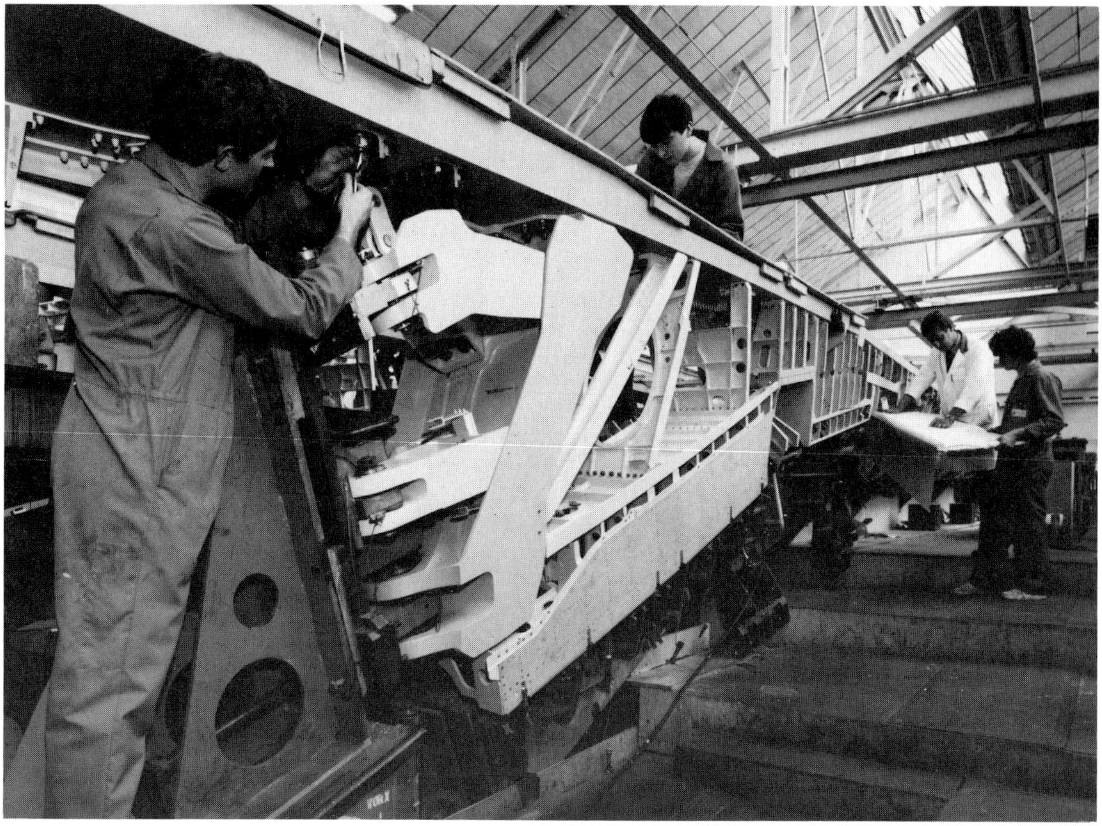

*Top:*
**Airways International Cymru is Britain's newest One-Eleven operator with two 300s, including G-YMRU.** *Airways International Cymru*

*Above:*
**Airbus wing assembly at the Filton, Bristol plant of British Aerospace.** *BAe*

*Below:*
***Sir Robert Watson-Watt*** **was the second A310 to be delivered, in March 1984. Sporting its French test registration in this picture, the aircraft received British registration G-BKWU.** *British Caledonian*

entered service with Air France on the Paris-London route on 30 May 1974.

This high-technology twin-engined wide-body airliner has since been produced in several variants with either General Electric or Pratt & Whitney engines. The latest version, the A300B4-600, has accommodation for up to 375 passengers.

A decision to proceed with the smaller capacity A310, seating 200 passengers, was made in July 1978, and the prototype (F-WZLH) first flew on 3 April 1982. It entered service with Swissair and Lufthansa in April 1983. The go-ahead for the new 150-seat A320 to complement the existing A300 and A310 versions was given in 1984. Designed as a next generation, single-aisle, twin-engined airliner, the A320 is being considered in two versions, the A320-100 to fill short/medium haul requirements, and the extended range A320-200 variant. Although it is not scheduled to enter airline service until 1988, six airlines have already placed orders for a total of 136 aircraft (66 firm orders and 70 options), the largest pre-launch order book of any of the Airbus family. The only UK airline to have ordered the A320 to date is British Caledonian with seven aircraft. First flight is set for February 1987. In terms of workshare, BAe will have an increased participation in the A320 of 26%, compared with approximately 20% in the other versions.

Although slow to get started, the Airbus in its two versions has now become firmly established in the world airline market. With over 400 firm sales for the A300/A310/A320 from 47 major airlines, accounting for 30% of the total market for wide-bodied airliners, the consortium has achieved greater success in the civil aircraft market than any previous European manufacturer.

## SHORTS

The history of the Shorts organisation dates from April 1901 when the brothers Oswald and Eustace Short set up business at Hove in Sussex for the manufacture of aerial balloons. Within two years they had transferred to larger premises in London and, after they were joined by their eldest brother, Horace, in 1908, they moved into the new era of powered flight. In 1909 the brothers were awarded a contract by Orville and Wilbur Wright for the manufacture of six biplanes under licence, which were constructed at Shell Beach on the Isle of Sheppey. In them, members of the Aero Club gained the first pilots' licences to be issued in the United Kingdom.

From these modest beginnings the company rose to become one of the most famous in British aviation, pioneering a long line of original designs which were as varied in their concept as they were in appearance, and included biplanes, flying-boats, heavy bombers, airships and the UK's first fixed-wing vertical take-off and landing aircraft. Names like Silver Streak, Scion, Singapore, Sunderland, Stirling, Solent and Sandringham recall at once the heyday of British manufacture and pay tribute still to the importance of the Short Brothers' contribution to it.

The headquarters of the Shorts organisation is now located at Belfast, in Northern Ireland, where the company operates a design and production complex which is one of the best equipped of its kind in Europe. A broad-based work programme is concentrated in three main areas — aircraft, aero-structures and missiles — employing 6,500 people. Aircraft activity covers the whole area of design, development and manufacture of the company's own aircraft projects, which on the civil side currently include the Skyvan utility transport and the 330 and 360 commuter aircraft. Major aerospace companies with which Shorts are collaborating include Boeing, Lockheed and Rohr in the USA, and Fokker, British Aerospace and Rolls-Royce in Europe. During last year, Shorts and Embraer of Brazil signed an agreement to co-operate on advanced commuter airliners, a field in which both companies are already among the leading manufacturers in the world.

Shorts re-entered the civil market (after the Sealand and Solent flying-boats in the postwar years) with the SC.7 Skyvan, a 19-seat STOL utility transport which first flew in 1963. In February 1970 it became the first aircraft to be certificated under the ARB's new Civil Airworthiness Requirements for STOL operations. A square, ugly looking aircraft, the Skyvan has nevertheless proved its worth with many different operators as a functional, reliable and economic aircraft and continues to serve throughout the world in a variety of civil and military transport roles. A total of 150 of all versions have been built to date and production is continuing with the Series 3 Skyvan. Whilst the Skyvan, in spite of high hopes, never achieved the commercial success it deserved, it provided enough encouragement for Shorts to persist with the same design philosophy when market research into the growing commuter business indicated a requirement for a similar but larger aircraft with increased seating for up to 30 passengers. The aircraft which finally emerged, its ancestry evident in the square fuselage, was the SD3-30, now known as the Shorts 330.

### Shorts 330

The Shorts 330 is a 30-passenger twin-turboprop

**It is flying-boats for which Shorts is perhaps best known among British manufacturers. (*Top*) Short Kent boats *Scipio*, *Sylvanus* and *Satyrus* and (*above*) S.45 Solent 3 G-AKNR.** *British Airways*

*Top:*
**This historic photograph was taken near Leysdown, on the Isle of Sheppey, in 1909, the year in which the Wright brothers placed a contract with Shorts for the manufacture of six biplanes. The group includes Oswald, Horace and Eustace Short (second, third and fourth, back row), and in the front row (from left) are J. T. C. Moore-Brabazon (later Lord Brabazon of Tara), Wilbur and Orville Wright and C. S. Rolls (co-founder of Rolls-Royce).** *Shorts*

*Above:*
**There is no Skyvan currently on the British register, although the type continues to give sterling service in other parts of the world. The most recent UK operator was Vernair of Liverpool which used G-AWWS for the transportation of race horses until 1983.** *Paul A. Tomlin*

*Above:*
**G-BDBS was the second prototype of the Shorts 330. It made its maiden flight on [illegible] July 1975.**
*Shorts*

airliner designed primarily for the commuter and regional airline market. The development programme was undertaken with two prototypes and the first production aircraft and spanned a total of over 1,000 flying hours over a period of 22 months. Metal for the first prototype was cut initially in August 1973 and the aircraft (G-BSBH) made its maiden flight on 22 August 1974. CAA certification was granted on 18 February 1976.

The first order was placed in August 1974 by Command Airways of Poughkeepsie, New York State and deliveries commenced in June 1976. The first revenue service was flown on 24 August 1976, only two years after the first flight, by the Canadian regional carrier Time Air, between Lethbridge, Calgary and Edmonton, Alberta. The first success in the British market came in 1979 when Loganair took delivery of the first of two aircraft.

Initial production aircraft, subsequently designated 330-100, were powered by the Pratt & Whitney Aircraft of Canada PT6A-45A, then -45B, engines but these were replaced by the more powerful PT6A-45R in the later 330-200 series which is now the standard passenger version. Apart from the engines there were also a number of detailed improvements as standard which on the earlier model were available only as options. The 330-200 combines large airliner style 'wide-body' comfort with small-aircraft economics. The 1.93m square cabin, designed by Walter Dorwin Teague Associates of New York, accommodates 30 passengers in 10 rows of three-abreast seating, plus cabin attendant. A standard forward cargo door facilitates the use of the aircraft in mixed passenger/cargo, or all-cargo configurations with a maximum payload of 3,400kg.

Other versions currently available are the 330-UTT, a military utility tactical transport which can carry up to 30 troops, 30 paratroops, or 15 stretchers and four seated patients or medical assistants for casualty evacuation, and the Sherpa, a freighter derivative of the 330-200 with a full-width, ramp-type, rear-loading door. The Sherpa got a big lift-off in the summer of last year when the US Air Force confirmed an order for 18 aircraft.

At 1 July 1985 a total of 167 firm orders and options had been placed for the 330 and variants of which 106 had been delivered. Production is continuing at 1½ aircraft per month.

### Shorts 360

The Shorts 360 is essentially a stretched development of the 30-seat 330, providing seating for six more passengers in a fuselage lengthened by the insertion of a 0.91m section forward of the wings. The other major visual difference is a redesigned tail unit which consists of a conventional, swept-back, single fin and rudder, replacing the twin assembly of the 330. The aircraft was designed specifically for short

| | **Shorts Skyvan 3** | **Shorts 330-200** | **Shorts 360** |
|---|---|---|---|
| *Dimensions* | | | |
| Wing span | 19.78m | 22.76m | 22.81m |
| Length | 12.20m | 17.69m | 21.59m |
| Height | 4.60m | 4.95m | 7.21m |
| Wing area | 35.11sq m | 42.08sq m | 42.10sq m |
| *Weights* | | | |
| Maximum take-off | 5,670kg | 10,387kg | 11,793kg |
| Maximum payload | 2,086kg | 3,400kg | 3,765kg |
| *Powerplant* | 2×533kW Garrett AiResearch TPE331-2-201A turboprops | 2×893kW PWAC PT6A-45R turboprops | 2×990kW PWAC PT6A-65R turboprops |
| *Performance* | | | |
| Maximum cruising speed | 324km/hr | 350km/hr | 393km/hr |
| Take-off field length | 610m | 1,188m | 1,362m |
| Range with max payload | 296km | 876km | 876km |
| Range with maximum fuel | 1,324km | 1,695km | 1,697km |
| Service ceiling | 6,860m | 3,500m | 3,500m |

***Top:***
**Shorts 360 36-seat airliners in production at Shorts' Belfast facility.** *Shorts*

***Above:***
**Shorts 360 G-BKMX is one of two operated by Loganair on its Scottish network.** *Shorts*

haul operations over typical commuter stage lengths of around 200-250km.

Details of the 360 were first announced by Shorts in July 1980, and the prototype (G-ROOM) made its first flight on 1 June 1981, some six months ahead of schedule and in time for the Paris Aero Salon. It was initially powered by PT6A-45 engines, but the more economical PT6A-65R powerplant, selected for the production run, was installed prior to CAA certification which was obtained on 3 September 1982. US Certification followed in November. The first production aircraft made its maiden flight on 19 August 1982 and entered service with Suburban Airlines of Reading, Pennsylvania on 1 December 1982. The first British airline to operate the 360 was Genair, and current operators include Air Eçosse, Air UK, Loganair, British Midland and Manx Airlines.

At 1 July 1985 firm orders and options totalled over 120 aircraft.

### Shorts 450

The Shorts 450 is a new 44/49-seat regional airliner development, unveiled in model form at the Paris Air Show in May 1985. If it is launched in the near future, the 450 could fly by the end of 1987.

## PILATUS BRITTEN-NORMAN

The partnership between Desmond Norman and the late John Britten began after both had completed their apprenticeships at the de Havilland Technical School. Their first project was an ultra-light monoplane, the BN-1 Finibee, which made its first flight on 26 May 1951 but was not put into production. Somewhat disappointed, they turned their attention to converting surplus training aircraft for agricultural use, in which they achieved much more success. After a flirtation with hovercraft, Britten and Norman came back to aircraft construction, and in 1964 work began in earnest on a new rugged, lightweight feeder transport. The result was the 10-seat twin-engined BN-2 Islander, which has been sold in greater numbers than any other commercial aircraft outside the United States, with total deliveries currently standing at over 1,050.

An attempt at a larger version, the three-engined Trislander with accommodation for 17 passengers, met with less success, with only 85 aircraft produced in the 11 years between 1971 and 1982. A four-seat design, the BN-3 Nymph, fell victim to the depressed state of the market in the early 1970s, and only one example was built.

In spite of the tremendous success with the Islander and a back-log of orders, the company experienced severe cash-flow problems and a receiver was called in. In August 1972 Britten-Norman (Bembridge) was finally acquired from the receiver by the Fairey Group, which set up a new production line at its Gosselies factory in Belgium, relegating the Bembridge facility to fitting out and certification work only.

Soon after, a feasibility study was carried out into a 100-seat transport, referred to as the Mainlander, which was to be a three-engined high-wing monoplane with Rolls-Royce Dart engines. This was far too ambitious a step, however, and the project was shelved.

1977 was a particularly difficult year. First came the untimely death of John Britten, one of the country's foremost designers, and this was followed soon after by the collapse of Fairey. In September 1979 Pilatus Aircraft Ltd of Stans in Switzerland acquired all assets of Britten-Norman (Bembridge) Ltd, including the facilities on the Isle of Wight and the former Fairey Islander/Trislander production hardware at Gosselies.

### Islander

The Islander is a 10-seat high-wing cantilever monoplane powered by two Lycoming O-540 or IO-540 piston engines. Conceived originally as a STOL feederline transport, the Islander's simplicity, reliability and versatility is its strength, especially in countries with limited airfield and maintenance facilities.

Detail design of the Islander was started in April 1964 and the prototype (G-ATCT) flew for the first time on 13 June 1965. Deliveries to Glos-Air and Loganair in the UK began three days after the type received its CofA on 10 August 1967. Initial production aircraft were designated BN-2, but those built from 1 June 1969 had the designation BN-2A, incorporating aerodynamic improvements as well as a side-loading baggage facility. This version was superseded in 1978 by the BN-2B which differs primarily in having a higher maximum landing weight, improved interior design and smaller propellers to reduce cabin noise levels. This current production version is available in several configurations, with a choice of three different powerplants — the Lycoming O-540 and IO-540 piston engines and the Allison 250-B17C turboprop (with the latter it is designated the BN-2T Turbine Islander) — and with or without auxiliary fuel tanks.

The Islander is also built in two military versions, the Defender and Maritime Defender, and the latest model, the CASTOR Islander ('Corps Airborne Stand-Off Radar'), a Turbine Islander with a distinctive nose radome, made its appearance at the 1984 Farnborough Air Show.

The Islander's many roles include passenger and freight transport, air ambulance duties,

***Above:***
**An aerial study of the BN-2T Turbine Islander**
***Pilatus Britten-Norman.*** **The new sliding door fitted to the same model.** *Pilatus Britten-Norman*

| | BN-2B Islander | BN-2T Turbine Islander | BN-2A Mk III-2 Trislander |
|---|---|---|---|
| *Dimensions* | | | |
| Wing span | 14.94m | 14.94m | 16.15m |
| Length | 10.86m | 10.86m | 15.01m |
| Height | 4.18m | 4.18m | 4.32m |
| Wing area | 30.19sq m | 30.19sq m | 31.31sq m |
| *Weights* | | | |
| Maximum take-off | 2,993kg | 3,175kg | 4,536kg |
| Maximum payload | 929kg | 1,113kg | 1,586kg |
| *Powerplant* | 2×194kW Avco Lycoming 0-540-E4C5E piston engines | 2×298kW Allison 250-B17C turbine engines | 3×194kW Avco Lycoming 0-540-E4C5 piston engines |
| *Performance* | | | |
| Maximum cruising speed | 257km/hr | 315km/hr | 267km/hr |
| Take-off field length | 360m | 381m | 550m |
| Range with max payload | 387km | 261km | 230km |
| Range with maximum fuel | 1,250km | 1,347km | 1,209km |
| Service ceiling | 4,450m | 7,620m | 4,010m |

agricultural spraying, aerial survey and photography, parachute training, oil pollution dispersal, maritime patrol and many others. As at 1 March 1985 over 1,050 Islanders of all types had been sold to more than 400 operators in 120 countries. This total includes 395 aircraft produced to date by IAv Bucuresti in Romania, and 67 aircraft built by the Philippine Aerospace Development Corporation between 1974 and 1980.

### Trislander

The Trislander is an enlarged version of the Islander with accommodation for up to 17 passengers. The major changes consists of a third engine in the tail unit, the insertion of a fuselage section forward of the wing, strengthening of the rear fuselage, and a new landing gear. With the same fuselage cross-section, much commonality was retained with the smaller Islander.

The second prototype of the Islander (G-ATWU) was converted to the prototype Trislander and flew for the first time on 11 September 1970. Certification was obtained on 14 May 1971, and the first aircraft was delivered to Aurigny Air Services in Alderney on 29 June 1971. The first production version was the BN-2A Mk III, and this was followed by the Mk III-2 with an optional extended nose section, and the Mk III-3 with automatic feathering of the propeller.

The Trislander was sold to many parts of the world where it still gives excellent service, but production ceased in 1982 after the completion of 73 aircraft. In addition, 12 aircraft are to be assembled under licence by International Aviation Corporation of Homestead, Florida, from British-built components and marketed as the Tri-Commutair.

## EDGLEY AIRCRAFT

Edgley Aircraft was originally formed in 1972 to carry out a design study for a light fixed-wing alternative to the helicopter, and in 1978 decided to produce a purpose-built observation aircraft, the EA7 Optica. Designed by John Edgley, this revolutionary aircraft made its first flight on 14 December 1979, generating a considerable amount of interest at various airshows during the subsequent years. Certification was received in February 1985, by which time the company had received orders for 80 aircraft, totalling £8million, from customers in 25 countries including the UK, continental Europe, Scandinavia, Australia, New Zealand, Saudi Arabia, Kenya, Tanzania and Brazil.

Production is undertaken at Edgley Aircraft's own modern computer-aided, 5,575sq m manufacturing facility at Old Sarum airfield near Salisbury in southern England, which was acquired from the Ministry of Defence in November 1982. An initial production run of 200 aircraft has been laid down, and the first delivery was made in April 1985 to Air Foyle of Luton. Other UK customers to date include Cabair and Northern Executive Aviation with three aircraft ordered by each. Present production rate is five aircraft per month, increasing to seven by 1986.

**Optica**

Everything about the Optica is startlingly original. Powered by the well-proven Avco Lycoming IO-360 engine driving a unique ducted fan, the Optica combines outstanding qualities of operational efficiency with quietness and an extremely low vibration level. A spacious cabin allows three-abreast seating with enough space all round for baggage, cameras, observation equipment and the like. Its 'insect eye' structure offers all round visibility and optical qualities which give pilot and observer a 270deg panoramic vision and allows an almost vertical line of sight to the side, and 45deg downwards straight ahead.

Optica's acquisition and operating costs are less than one-third those of a helicopter offering comparable performance and visibility. Unlike a helicopter, however, Optical offers low noise and vibration, and a flight duration of up to seven hours including reserves. The Optica also has very good slow-flight characteristics, a low stalling speed, and STOL capability from both hard and soft airstrips. Maximum speed is 222km/hr. Roles for which the Optica is particularly suited include aerial photography, film, television and press reporting, security patrols, traffic watch and other police duties, pipeline and powerline inspection, forestry patrol, archaeological exploration, coastguard duties, tourist sightseeing flights, agricultural observation, game spotting, mineral ore prospecting and geological and geophysical survey work.

These are only some of the roles currently envisaged. There is no doubt that many others will be added to this list once the Optica has accumulated some full-scale service hours.

---

**Edgley EA7 Optica**

*Dimensions*
Wing span: 12.00m
Length: 8.16m
Height: 2.31m
Wing area: 15.84sq m

*Weights*
Maximum ramp weight: 1,236kg
Useful payload: 355kg

*Powerplant*
One ducted propulsor unit with 194kW Avco Lycoming IO-540

*Performance*
Maximum level flight speed: 222km/hr
Loiter observation speed: 122km/hr
Endurance with normal reserves: 6-9hr
Maximum range with normal reserves: 1,139km
Service ceiling: 4,115m

---

## ROLLS-ROYCE

Rolls-Royce is the largest manufacturer of aero gas turbines in Europe and ranks as one of the Western world's three major aero-engine companies. Now wholly government owned, the company employs a total of 42,000 people in seven locations in the UK, engaged in the development and production of a wide range of gas turbines for civil and military aircraft. Derivatives of Rolls-Royce aero engines are also in service worldwide for the propulsion of ships and hovercraft, and for electrical power generation and oil and gas pumping and compression. Some 2,300 engines for industrial and marine purposes sold to 57 countries, and powerplants for aircraft in operation with 300 airlines, 112 armed forces and 670 corporate operators, makes impressive reading.

Rolls-Royce introduced the world's first turbofan engine, and continued development of the revolutionary vectored-thrust engine, the Bristol-Siddeley Pegasus, for vertical take-off and landing, as installed in the Harrier. The company continues in the forefront of new-technology development with highly-advanced, quieter and more fuel-efficient engines, and is also involved in a number of collaborative ventures with engine manufacturers in Germany, France, Italy, Japan and the United States. Working associations on design, development and production are maintained with a total of 15 countries.

**Turbofans and Turbojets**

Current turbofans in production for civil use include the RB211, installed in the Lockheed TriStar, Boeing 747 and 757, and the Spey, which powers the One-Eleven, Trident, Fokker Fellowship and the Gulfstream II and III. The **RB211** is an advanced three-shaft turbofan which entered service in 1972, and has accumulated over 14 million hours flight time. Versions produced to date have a thrust range from 166.3kN to 235.8kN; over 1,300 have been delivered.

The **Spey** first entered service in 1964, and more than 4,700 units have been produced since that date. It powers a total of nine civil and military aircraft, and civil versions produce a thrust range of 43.8kN to 55.8kN.

Being developed by **International Aero Engines**, a specially-formed consortium of

# Optica

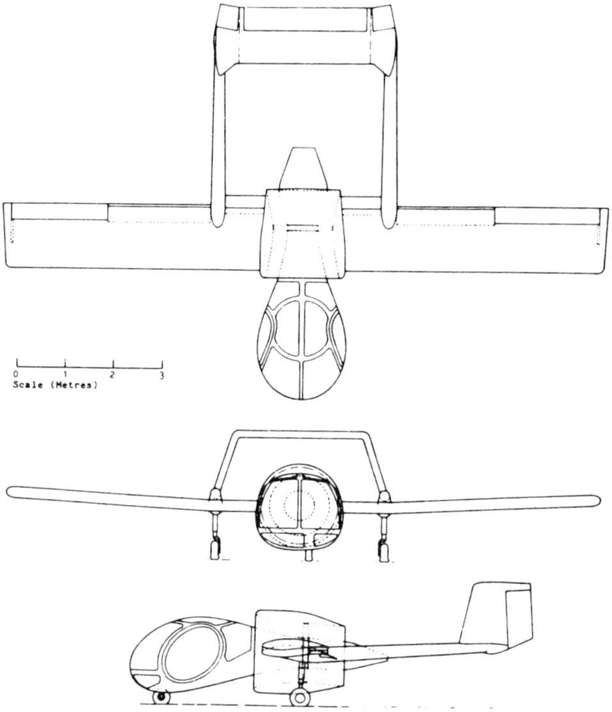

***Above:***
**The Flight of the Bumblebee. An Optica flying past the Needles off the Isle of Wight.** *Edgley*

Rolls-Royce, United Technologies, MTU, Japanese Aero Engines and Fiat, is the **V2500**, an advanced technology turbofan aimed at the new generation 150-180 passenger airliners. It is expected to enter service in 1989. The **RB401** with a thrust of 24.6kN is an advanced turbofan under development for the business jet and light trainer market, and the **Tay**, a 59.7kN twin-shaft turbofan designed to meet the demands of new short/medium haul airliners and longer range business aircraft, is scheduled for certification in July 1986. Three hundred units of the latter have been ordered for the Fokker F100 and the Gulfstream IV.

The largest turbojet is the 169kN thrust **Olympus** produced jointly with SNECMA of France for the Concorde supersonic airliner. It was the first engine to achieve a full passenger-carrying certificate for supersonic operations.

The **Viper** is a small engine providing up to 17.8kN thrust. Over 5,500 have been sold to date and are installed in many military aircraft, as well as the BAe 125 and Piaggio PD-808 business jets.

### Turboprops and Turboshafts

The company's most famous turboprop is the **Dart**, which was the world's first turboprop to enter airline service with the Vickers Viscount in 1953. Since then it has accumulated over 100 million hours of operation. Nearly 7,000 Darts have been produced, covering a power range from 1,150 to 2,420kW, and the engine continues in production for the BAe 748 and Fokker Friendship airliners. Dart engines also power the NAMC YS-11, Gulfstream I and Argosy.

The **Tyne** is a large turboprop which delivers up to 4,750kW. It powers the Vickers Vanguard, Canadair CL-44, Short Belfast, Breguet Atlantic, Aeritalia G-222 and Transall C-160 civil aircraft. Over 1,200 engines have been built.

Rolls-Royce has also produced several helicopter engines for both civil and military use, including the Gnome and Gem turboshafts. The 1,238kW **Gnome** entered service in 1961, and continues in production for the Westland Sea King and Commando helicopters. It also powers the Whirlwind and Wessex. Over 2,000 engines have now been delivered. The **Gem** is a 835kW twin-spool turboshaft which is installed in the new Westland 30 civil helicopter, as well as the Westland Lynx and Agusta A129. More than 600 have been delivered since it entered service in 1974. The **RTM322** is under development for the EH 101, a joint project between Westland and Agusta of Italy.

*ADDRESSES*

**British Aerospace plc**
100 Pall Mall
London SW1Y 5HR

**British Aerospace plc**
Civil Division
Hatfield
Hertfordshire AL10 9TL

and at Chester, Filton and Prestwick

**British Aerospace plc**
Warton Division
Warton Aerodrome
Preston
Lancashire PR4 1AE

and at Preston and Samlesbury

**British Aerospace plc**
Weybridge Division
Brooklands Road
Weybridge
Surrey KT13 0SF

and at Dunsfold, Hamble, Kingston, Brough and Woodford

**Edgley Aircraft Ltd**
Old Sarum Airfield
Salisbury
Wiltshire SP4 6BN

**Pilatus Britten-Norman Ltd**
Bembridge Airport
Bembridge
Isle of Wight PO35 5PR

**Short Brothers Ltd**
PO Box 241
Airport Road
Belfast
Northern Ireland BT3 9TZ

**Rolls-Royce Ltd**
65 Buckingham Gate
London SW1E 6AT

**Rolls-Royce Ltd**
Aero Applications
PO Box 31
Derby
Derbyshire DE2 8BJ
*and*
PO Box 3
Filton
Bristol BS12 7QE

**Rolls-Royce Ltd**
Helicopter Applications
Leavesden
Watford
Hertfordshire WD2 7BZ

and at Barnoldswick, Glasgow, Coventry and Sunderland

**Society of British Aerospace Companies Ltd**
29 King Street
St James's
London SW1Y 6RD

## ROUND-UP

A number of British aircraft now out of production and/or manufactured by companies no longer in existence still give valuable service throughout the world. These pages are intended to reach those parts other aeroplanes cannot reach.

### Armstrong Whitworth AW.650 Argosy

This distinctive high-wing twin-tail-boom cargo aircraft evolved from design studies initiated by Armstrong Whitworth in 1955. Powered by four Rolls-Royce Dart turboprops, the first aircraft (G-AOZZ), a Series 100, made its maiden flight on 8 January 1959. Ten of these were followed by seven Series 220 (first flight 11 March 1964), which, though externally very similar, differed considerably in detail.

In addition to the civil Argosies, 56 military C.Mk 1s were built for the RAF. Seven remain in service, three with IPEC in Australia, two with Safe Air and two with Air Bridge Carriers in the UK.

### Bristol 170 Freighter

The Bristol Freighter is a twin-engined high-wing utility transport powered by two Bristol Hercules piston engines. It was built in two main versions, the Mk I freighter, and the Mk II Wayfarer for up to 32 passengers. The freighter prototype (G-AGPV) first flew on 2 December 1945, followed by the Wayfarer (G-AGVB) on 30 April 1946. The Bristol 170 gained fame in the 1950s when used on the cross-Channel passenger and car ferry services, and in later years on the Wellington-Chatham Island run by Safe Air. When production ended in 1958, a total of 214 of all variants had been built. A few are still in service in New Zealand with Safe Air and Hercules Airlines, and one in the UK with Atlantic Air Transport.

### Bristol 175 Britannia

Known as the 'Whispering Giant', the Britannia is a four-engined medium/long range turboprop whose early success was cut short by the advent of jet aircraft. The first prototype (G-ALBO) made its maiden flight on 16 August 1952, and two production versions followed, the Series 100 with accommodation for up to 90 passengers, and the Series 300, an enlarged version for 133 passengers. Both were powered by Bristol Proteus turboprops.

Production totalling 85 aircraft ended in 1960. Two derivatives, the Canadair CL-28 Argus maritime reconnaissance aircraft, and the CL-44 transport, were built in Canada. Aero Caribbean of Cuba operates four Series 300, and four ex-RAF machines have found their way to Zaire.

### de Havilland DH.114 Heron

A four-engined companion to the Dove, the prototype Heron (G-ALZL) flew for the first time on 10 May 1950, powered by Gipsy Queen engines. Initial production version was the Heron 1 with fixed undercarriage, and this was replaced by the Heron 2 with retractable undercarriage from 1954.

Although capable of carrying up to 17 passengers, the Heron also became very popular as a VIP transport. By the time production ended, a total of 148 Herons had been built, these being 51 Series 1 and 97 Series 2. Many subsequent conversions gave the Heron a new lease of life, including the Riley Turbo Skyliner, and the Saunders ST-27, a greatly modified version for 23 passengers.

Almost half of all Herons built continue in service, but the sole UK operator is Topflight Aviation at Fairoaks.

### Handley Page HPR.7 Herald

The Herald is a high-wing, twin-engined medium range airliner powered by two Rolls-Royce Dart 527 turboprops. Designed by Handley Page's Reading division (formerly the Miles Aircraft Co), the first prototype (G-AODE) made its first flight on 25 August 1955. Production totalled 50 aircraft made up of four 47-seat Series 100, 38 56-seat Series 200, and eight Series 400 (a military version for the Malaysian Air Force). Production ceased with the collapse of Handley Page in 1970.

Twelve are still in service in the UK, Italy, Zaire and Colombia. Heralds in Britain include two with BAF, four with Channel Express, and one each with Elan Air, Janus Airways and Skyguard.

### Hawker Siddeley HS.121 Trident

The Trident originated as a de Havilland proposal to meet a BEA requirement for a fast short/medium haul transport. A rear-engined jet airliner powered by three Rolls-Royce Spey

*Top:*
**Afrek's ex-Royal Air Force Britannia, G-BDUR, was the last British-registered aircraft in commercial service.**

*Above:*
**With Topflight, the Heron is available for charter in a variety of configurations, including eight-seat executive or 14/15-seat commuter layouts, or in all-freight with a capacity of 1.5 tonnes.** *Topflight*

turbofans, the first aircraft, a Trident 1 (G-ARPA), made its maiden flight on 9 January 1962. Other variants followed including the 1E, 2E and 3B covering a seating capacity from 103 to 180 passengers.

From the outset the Trident was designed to have a Smiths Autoland system, and certification for automatic landing in CAT III weather conditions was obtained in September 1968.

A total of 117 aircraft was produced between 1962 and 1975, consisting of 24 Trident 1, 15 1E, 50 2E, 26 3B and two Super 3B, and some 20 are still in airline service with CAAC (China) and Air Charter Service in Zaire.

**Shorts SC.5 Belfast**

The Belfast was built as a heavy military transport for the RAF, able to carry the bulkiest military equipment then in use, or up to 250 troops. A high-wing aircraft with four wing-mounted Rolls-Royce Tyne turboprops, the Belfast first flew on 5 January 1964, and entered service with the RAF on 20 January 1966. Only 10 aircraft were built.

The Belfast has taken on a new lease of life after being converted for civil use for HeavyLift Cargo Airlines by Marshall's of Cambridge. This Stansted-based operator utilises three aircraft for the carriage of outsize and specialised freight.

**Vickers V.700/800 Viscount**

The Viscount, a 40-71 seat short/medium haul transport powered by four Rolls-Royce Darts, was the world's first turboprop airliner, bringing new speed and comfort to the air and doing much to help BEA become a major airline in the 1950s.

The first aircraft (G-AHRF), a Series 630, took to the air on 16 July 1948, but was soon superseded by the main production Series 700 which first flew on 19 April 1950. An extended fuselage variant, the Series 800, was added in July 1956. The Viscount was bought by a number of major carriers, and has seen service in over 60 countries in all parts of the globe. Production totalled 440 aircraft, and some 50 remain in service. Almost half of these can be found in the UK with BAF, BMA, Euroair, Guernsey Airlines, London European Airways and Manx Airlines.

**Vickers V.950 Vanguard**

The Vanguard is a short/medium range airliner built as a Viscount replacement to BEA and Trans-Canada Air Lines requirements, and these two airlines were the only original customers. It unfortunately entered service at a time when most likely customers were already looking towards jet aircraft, and only 43 of this excellent aircraft were produced. Powered by four Rolls-Royce Tyne turboprops, the Vanguard has space for a maximum of 139 passengers, or can carry up to 20 tonnes of freight. The first aircraft (G-AOYW) made its maiden flight on 20 January 1959.

Of the 43 aircraft built, eight are still in service including five Merchantman freighter conversions with Air Bridge Carriers, two with Air Gabon, and one with Merpati Nusantara in Indonesia.

***Below:***
**The Trident 3B will soon disappear from the UK. British Airways is due to have withdrawn the remaining examples from its fleet by the end of 1985.** *British Airways*

***Below:***
**British Air Ferries is now the world's largest Viscount operator. The illustrated aircraft, G-AOYO, has served in the past with BEA, BKS Air Transport, Northeast Airlines and British Airways.**
*BAF*

***Top:***
**Argosy G-APRL in Elan markings photographed at East Midlands. The aircraft is operated by Air Bridge Carriers.**
*P. J. Cooper via Aviation Data Centre*

***Above:***
**G-APEP is one of five freighter conversions of the Vanguard operated by Air Bridge Carriers at East Midlands Airport.** *EMA*

# British Airships – Is there a future in the past?

When the British airship seemingly plunged into history with the disastrous crash of the publicly-funded R101 in 1930, it would have taken a brave man indeed to predict that 50 years later an airship would share the skies with supersonic jet aircraft. A rich man's fancy to indulge his boyhood memories? Far from it. Today's sophisticated and ultra-safe machines are the work of a naval architect turned self-taught airship designer called Roger Munk, who was convinced that with modern construction techniques and advanced avionics he could turn the airship into a useful form of transport, filling a variety of roles including the carriage of passengers and freight. He was lucky in finding Andrew Millar, a business consultant and self-confessed 'aviation nut', who was prepared to back him, but it does not take too much imagination to realise that the early years were fraught with difficulties. The hardest part of all was to convince the cynics that building a sedate 90km/hr airship in today's supersonic jet age was more than just a crazy idea.

Succeed they did against all the odds, and on a grey February morning in 1979 a pre-production prototype (G-BECE) made a faultless maiden flight from Cardington in Bedfordshire. The airship had returned to its home after half a century in the wilderness.

The first production 'Skyship 500' (G-BIHN) made its first flight two years later. The craft is owned by the government which paid £1.25million in a deal with Airship Industries, leasing it back for test and demonstration flying. This was the first time that government money has been put into airships since the 1920s.

After the formation of Imperial Airways in 1924, the British government initiated a research programme for the construction of large rigid airships for commercial operations, and in particular for the possible use on the 'Empire' routes. The outcome of this project was the simultaneous construction of two similar airships, the R100 by the Airship Guarantee Co (a subsidiary of Vickers), and the R101, designed and built by the Air Ministry Establishment at Cardington. Both airships first flew in 1929, the R101 (G-FAAW) on 8 October, followed by the R100 (G-FAAV) on 16 December. The R100 subsequently made nine flights totalling 297 hours, including a successful return flight to Montreal. The R101's 12 flights totalled only 127 hours.

On 4 October 1930, with a temporary Certificate of Airworthiness, the R101 slipped its mast at Cardington at the beginning of a flight to India. The next day it dived into the ground and burst into flames at Beauvais in northern France, killing 48 of the 54 persons on board. A public enquiry concluded that the main cause of the disaster was a large tear in the forward gasbag, resulting in loss of height followed by an explosion of the escaping hydrogen upon impact with the ground. The R101 crash proved to be the end of British interest in the development of airships for commercial purposes. The R100 was broken up shortly afterwards and sold for scrap.

**Airship Industries**

Airship Industries, now 81.4% owned by Bond Corporation Holdings of Perth, Western Australia, was incorporated in 1978, initially to develop a rigid design concept known as the Thermo-Skyship. By 1980 all development under Chief Designer Roger Munk was concentrated on the non-rigid programme culminating in September 1981 in the successful maiden flight of the first production airship, the Skyship 500. This ship, registered G-BIHN, undertook many certification and development trials, including daily flights at Farnborough in 1982, shuttle trials between Charles de Gaulle and Orly Airports in Paris, concept tourist operations from London's Dockland to the Palace of Westminster, and evaluation trials for coastal surveillance in the South of France.

The successful PACE trials (Patrol Airship Concept Evaluation) by the US Navy, Coast Guard and NASA in 1983, which led to an invitation to bid for a full evaluation including a 5-12 month lease and crew training, was the start of one of the most significant periods for Airship Industries. Other highlights followed in 1984 with the use of a Skyship 500 for live broadcast of the Los Angeles Olympic Games, the first aerial advertising contract from Fuji Film, and the first outright sale to Nikko Trading Co, a subsidiary of Japan Air Lines, which will use the Skyship for promotional purposes. The Skyship 500 was

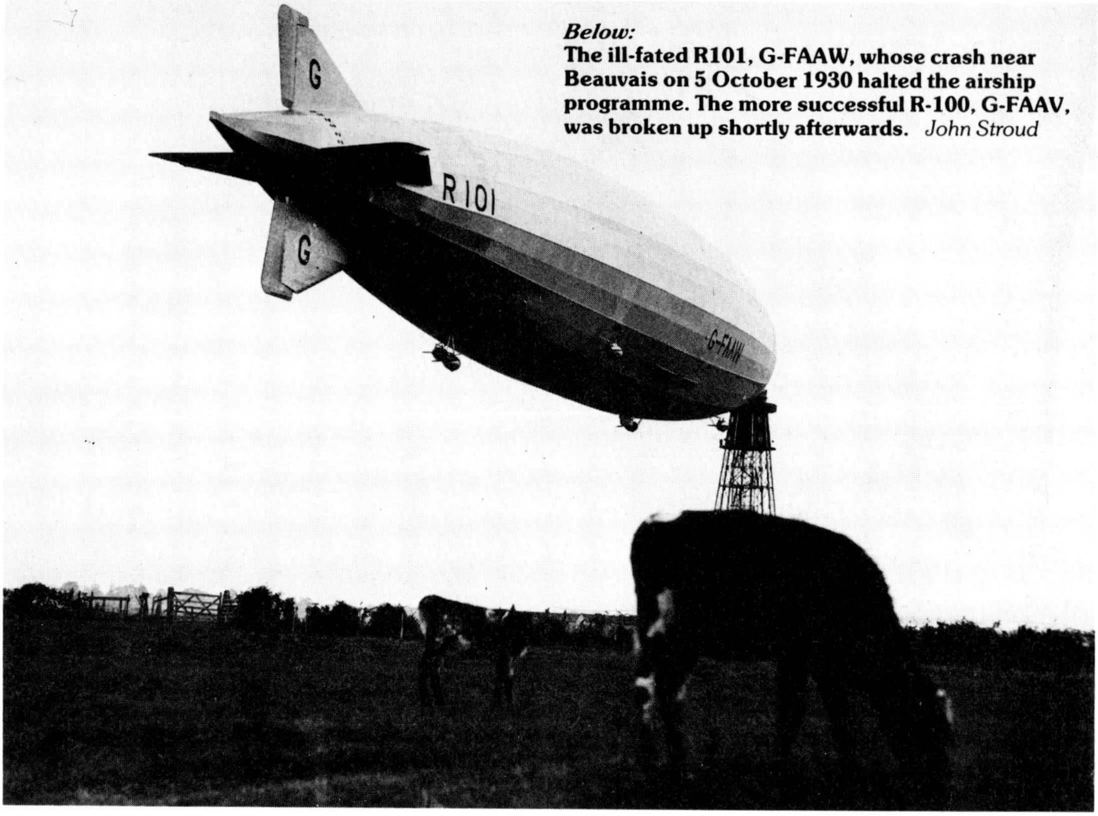


***Below:***
**The ill-fated R101, G-FAAW, whose crash near Beauvais on 5 October 1930 halted the airship programme. The more successful R-100, G-FAAV, was broken up shortly afterwards.** *John Stroud*

***Top:***
**Skyship 500 G-BIHN approaching the famous hangars at Cardington, Bedfordshire, home of the British airship.** *Airship Industries*

***Above:***
**Close-up of the gondola on the Skyship 500. The propulsion units are attached to the rear.**

fully certificated in November 1984, which made it the world's only airship to be certified to carry fare-paying passengers.

The larger Skyship 600 (G-SKSC) made its maiden flight on 6 March 1984, and full certification is expected in late 1985. The US Navy and Coast Guard, the British Ministry of Defence and the French Navy are all interested in the Skyship 600 and there have been enquiries from America, Canada, South Africa, China, Korea, Indonesia and Japan, In November 1984 the Skyship 600 joined the French Navy on successful fishery protection and shipping control trials off the Brittany coast, and the larger Skyship is also being considered by Interport for use on inter-island tourist shuttle services in the Aegean Sea.

With the research and development of the Skyships almost complete, the company is looking forward to a growing interest and more definite sales, which should also open the way for the much larger Skyship 5000 series.

**The Skyships**

The **Skyship 500** is a conventional, non-rigid airship with the payload compartment, known as the gondola, suspended from the helium gas-filled hull. Power comes from two 152kW air-cooled Porsche piston engines, mounted transversely inboard the rear of the gondola, driving unique ducted propellers which swivel to give vectored thrust for vertical take-off and landing. It can climb at 700m/min and has a maximum speed of 90km/hr. The interior of the gondola can be laid out so as to provide comfortable seating for up to eight passengers in addition to the two flight crew.

The **Skyship 600** is essentially a stretched version of the Skyship 500, with improved performance through the installation of turbocharged Porsche engines. A 30% increase in volume results in a 50% increase in payload, permitting greater comfort and long endurance, which makes it particularly suitable as a maritime patrol ship. In the civil passenger format the Skyship 600 can seat up to 18 passengers. It will also be the first aircraft ever to be fitted with fibre optics as a primary flight control system.

The **Skyship 5000** is a large derivative of the Skyship 500 intended for the US Navy's Battle Surveillance Airship System (BSAS), the UK Ministry of Defence OPV-3 (Offshore Patrol Vessel) requirement, and for inter-city passenger transport over relatively short ranges. Up to 200 passengers could be accommodated in first class comfort in a wide-body, two-deck layout, or in 80 cabins to bring 'ocean liner' comfort to the air. The lower deck could also be converted for the carriage of cargo. The Skyship 5000 could fly by 1988/89.

Apart from military applications where the airship's economy and great endurance is ideally suited to such tasks as surveillance, anti-submarine warfare, airborne early warning and many others, civil uses break down into four major categories: promotional, passenger, scientific and cargo. The promotional and passenger applications will often be combined, as the large envelope area of a passenger ship has considerable potential as aerial advertising space. The

| | **Skyship 500** | **Skyship 600** | **Skyship 5000** |
|---|---|---|---|
| *Dimensions* | | | |
| Gross envelope volume | 5,153cu m | 6,666cu m | 50,000cu m |
| Length | 52,0m | 59.0m | 108.0m |
| Diameter | 14.0m | 15.2m | 30.0m |
| Height | 18.7m | 20.3m | 41.0m |
| *Weights* | | | |
| Maximum weight | 5,250kg | 6,630kg | 50,000kg |
| Nominal disposable load | 1,965kg | 2,966kg | 30,000kg |
| *Powerplant* | | | |
| | 2×152kW Porsche 930/10 petrol engines | 2×200kW Porsche 930/67 petrol engines | 2×1,500kW PWAC PT7 turboprops |
| *Performance* | | | |
| Maximum speed | 90km/hr | 120km/hr | 170km/hr |

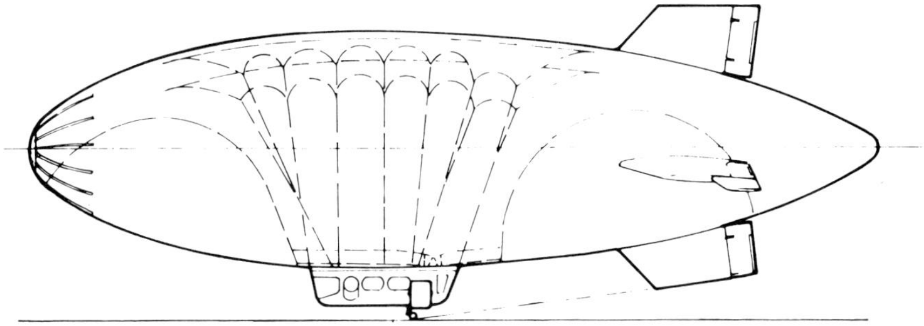

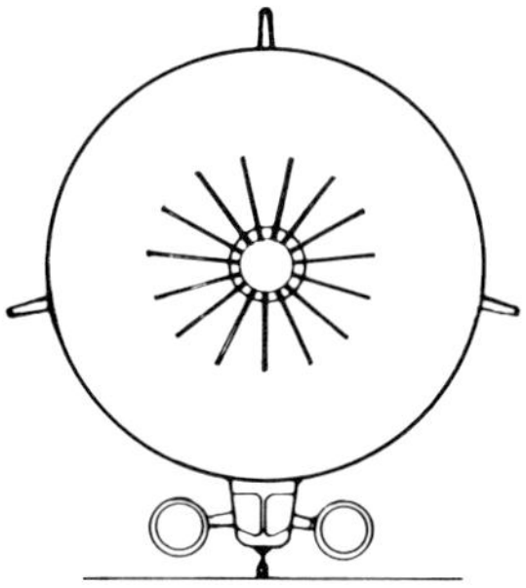

| Principal Dimensions | | |
|---|---|---|
| Gross envelope volume | 5153 m³ | 182,000 ft³ |
| Ballonet volume | 26% of gross envelope volume | |
| Length (overall) | 52.0 m | 170.6 ft |
| Diameter | 14.0 m | 45.9 ft |
| Height (overall) | 18.7 m | 61.2 ft |
| Tailspan | 17.0 m | 55.7 ft |
| L/D ratio | 3.71 | |
| **Gondola Dimensions** | | |
| Length (overall) | 9.24 m | 30.3 ft |
| Width (overall) | 2.41 m | 7.9 ft |
| Main cabin headroom | 1.96 m | 6.4 ft |
| Main cabin length | 4.20 m | 13.8 ft |

## SKYSHIP 500 GENERAL ARRANGEMENT

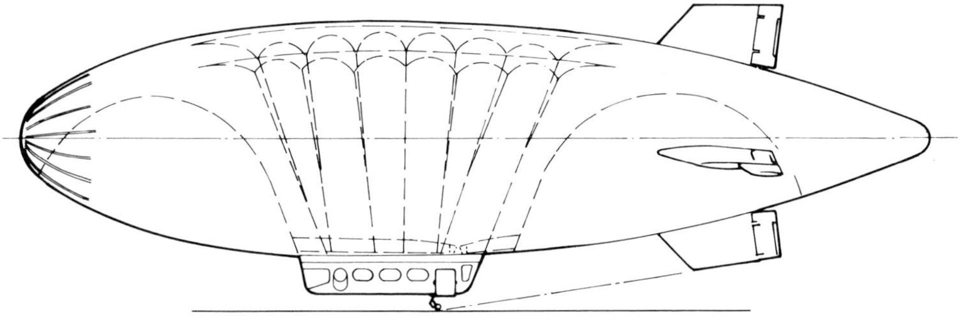

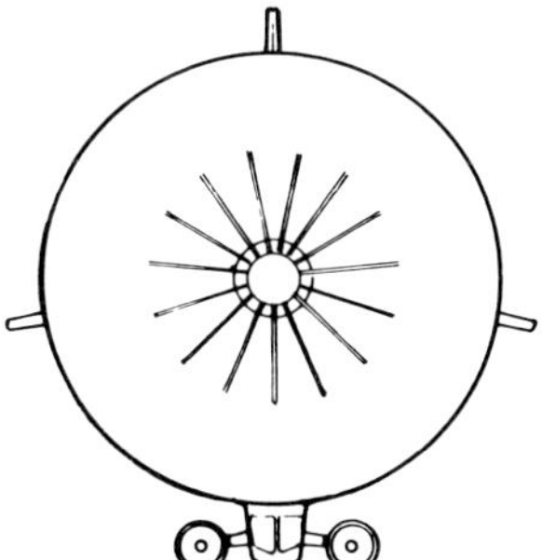

| Principal Dimensions | | |
|---|---|---|
| Gross envelope volume | 6666 m³ | 235,400 ft³ |
| Ballonet volume | 26% of gross envelope volume | |
| Length (overall) | 59.0 m | 193.6 ft |
| Diameter | 15.2 m | 49.9 ft |
| Height (overall) | 20.3 m | 66.6 ft |
| Tailspan | 19.2 m | 63.0 ft |
| L/D ratio | 3.88 | |
| **Gondola Dimensions** | | |
| Length (overall) | 11.67 m | 38.3 ft |
| Width (overall) | 2.56 m | 8.4 ft |
| Main cabin headroom | 1.92 m | 6.3 ft |
| Main cabin length | 6.89 m | 22.6 ft |

## SKYSHIP 600 GENERAL ARRANGEMENT

# SKYSHIP 5000 GENERAL ARRANGEMENT

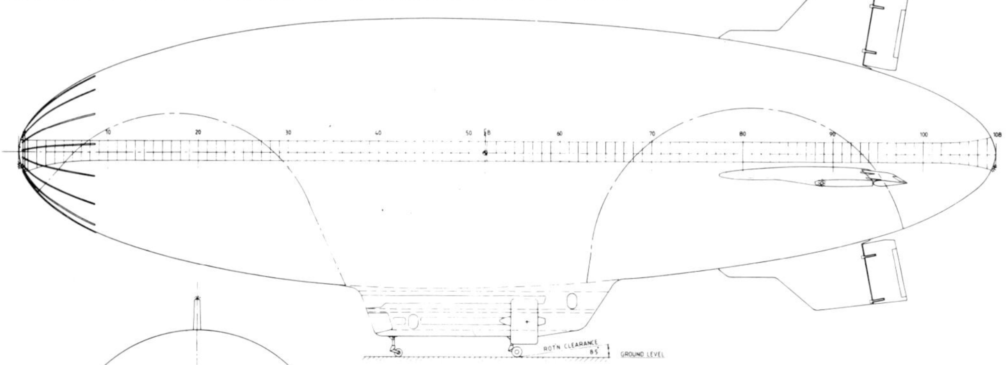


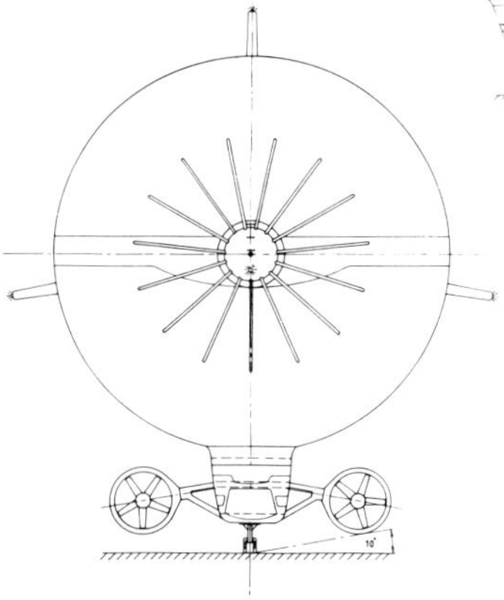


| | | |
|---|---|---|
| **Principal Dimensions** | | |
| Gross envelope volume | 50000 $m^3$ | 1,760,000 $ft^3$ |
| Ballonet volume | 32% of gross envelope volume | |
| Length (overall) | 108.0 m | 354 ft |
| Diameter | 30.0 m | 98 ft |
| Height (overall) | 41.0 m | 135 ft |
| Tailspan | 36.7 m | 120 ft |
| L/D ratio | 3.6 | |
| **Principal Weights and Loads** | | |
| Manufacturers Empty Weight | 20172 kg | 19.86 tons |
| Maximum Weight | 50000 kg | 49.22 tons |
| Nominal Disposable Load | 30000 kg | 29.53 tons |

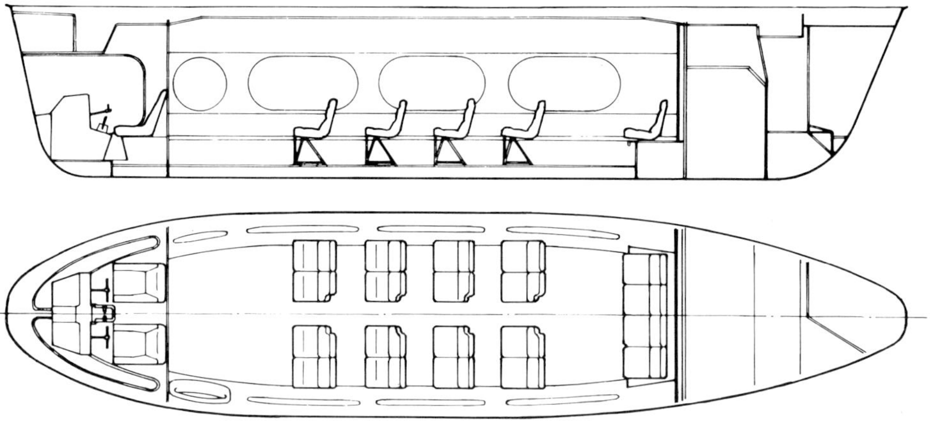

# SKYSHIP 600 GONDOLA ARRANGEMENT

***Below:***
**The airship over the B-1 bomber at Farnborough '82.** *Airship Industries*

scientific applications will include all types of surveys, especially geophysical and hydrographical, exploration and patrol work, but cargo usage is likely to be limited until the introduction of large payload airships such as the Skyship 5000.

**Production**

| | | | |
|---|---|---|---|
| 1214/01 | G-BECE | Prototype | First flown 3 February 1979 |
| 1214/02 | G-BIHN | SKS-500 | |
| 1214/03 | G-SKSA | SKS-500 | |
| 1214/04 | G-SKSB | SKS-500 | |
| 1214/05 | G-SKSE | SKS-500 | Nikko Trading Co |
| 1215/01 | G-SKSC | SKS-600 | First flown 6 March 1984 |
| 1215/02 | G-SKSD | SKS-600 | |
| 1215/03 | G-SKSG | SKS-600 | |
| 1215/04 | G-SKSF | SKS-600 | |
| 1215/05 | G-SKSJ | SKS-600 | |

**The Airship Association**

Founded in 1971, the Airship Association provides a forum for like-minded people who believe that the airship has an important role to play in the transport needs of the future. The Association holds symposia, conferences and lectures and maintains contacts with organisations and individuals in the UK and abroad and works closely with the Parliamentary Airship Group.

*ADDRESSES*

**Airship Industries (UK) Ltd**
84-86 Baker Street, London W1M 1DL

**The Airship Association Ltd**
3 Chestnut Avenue, London SW14 8NT

***Above:***
**Skyship 600, G-SKSC, photographed over London's Tower Bridge.**

AIRSHIP INDUSTRIES LTD.

# British Airways – Supremo of the sky

**British Airways** is Britain's national flag-carrier operating international and domestic scheduled and charter air services for the carriage of passengers, cargo and mail. As far as international routes are concerned, it is designated to exercise traffic rights held by the British government under Air Services Agreements made with other countries.

British Airways' wholly-owned subsidiary, **British Airtours Ltd**, carries out substantial charter operations based on Gatwick Airport.

**British Airways Helicopters Ltd** provides a wide range of services for the offshore oil industry worldwide, a search and rescue operation covering the northern North Sea from Sumburgh in the Shetlands, as well as operating a scheduled passenger service between Cornwall and the Isles of Scilly. Bases are maintained at Penzance, Aberdeen, Beccles and Sumburgh, with headquarters at Aberdeen.

In addition, there are two other subsidiaries engaged in aviation-related but non-flying, activities. **British Airways Engine Overhaul Ltd** undertakes the overhaul of aircraft engines and modules for British Airways and for other customers. Its two plants in Treforest, South Wales are equipped to handle nine types of aero engine, made by Rolls-Royce and Pratt & Whitney. About half of the work undertaken is for British Airways, with the remainder for customers mainly in the Middle and Far East. **British Airways Associated Companies Ltd** controls extensive worldwide hotel interests and the airline's investments in Air Mauritius and GB Airways, formerly Gibraltar Airways.

British Airways' principal place of business is at London's Heathrow Airport.

The airline came into being on 1 April 1972 when the British Airways Board assumed control and ownership of the British Overseas Airways Corporation and the British European Airways Corporation. On 1 April 1984 all assets and liabilities of the British Airways Board were vested in British Airways plc, in exchange for 180 million shares of £1 each issued to the Secretary of State for transport. This was done in preparation for privatisation, now likely in early 1986, when shares will be offered for sale to the public.

## Organisation

Since its foundation in 1972 the airline has undergone many organisational changes. The latest of these was put into effect in July 1983 when the previous three major service divisions — European, Intercontinental and Gatwick — were reconstituted under a Marketing Division into eight market centres and three business centres. The market centres comprise United Kingdom, North Europe, South Europe, Americas, Africa, Middle East/Indian subcontinent, Far East and Australia/New Zealand, and the business centres are Cargo, Charters and Tours. Each of these centres is separately accountable for its own profitability.

All centres are served by an Operations Division which incorporates all the direct airline support activities, including pilots, cabin crew, engineering and catering, as well as the actual ground operations. Marketing and Operations are in turn supported by Air Safety, Finance, Human Resources, Information Management, Legal and Secretarial, Marketplace Performance, Medical, Planning, Public Affairs, Security and Training.

## Scheduled Services

On the domestic front, the launch of the improved 'Super Shuttle' service on 30 August 1983 on domestic routes between London and Glasgow, Edinburgh, Belfast and Manchester has helped British Airways to regain some of the market share lost with the introduction of a competitive service by British Midland Airways in 1982. The Shuttle concept, which was originally introduction into British Airways in 1975, provides a guaranteed seat, without prior reservation, to any full fare passenger who reports at the stated time. This entails a readiness to operate a back-up aircraft to carry those full fare passengers, however few, who cannot be accommodated on the first aircraft.

Regional services are strengthened, especially at Birmingham and Manchester. British Airways has the largest network of scheduled services from Manchester as well as significant charter activities and a major handling business at the

## BRITISH AIRWAYS DOMESTIC NETWORK

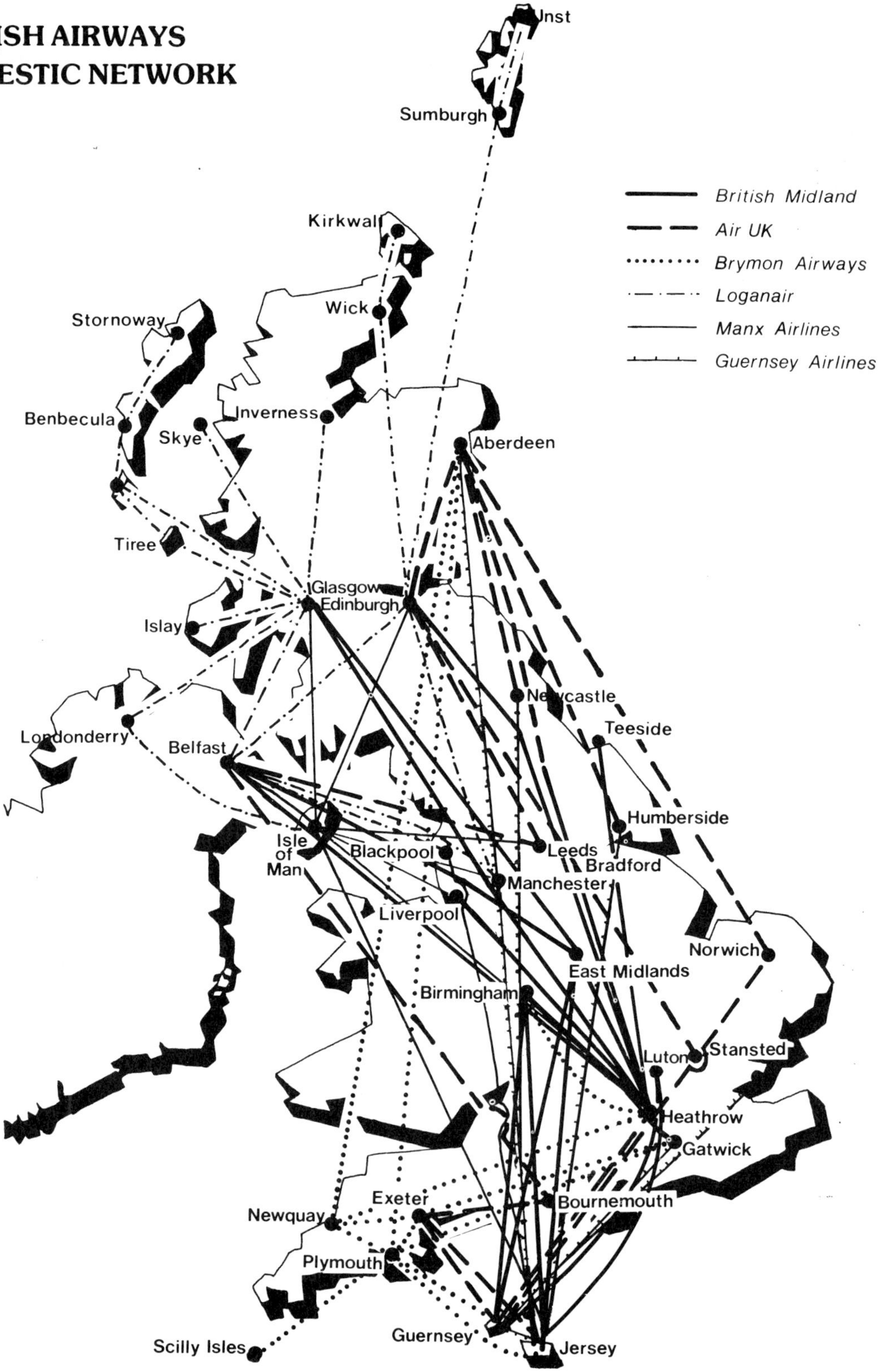

***Above:***
**The large and small end of British Airways' Boeing range includes Boeing 747 G-AWNO and Boeing 737-200 G-BKYF.** *British Airways*

***Above:***
**Looking splendid in their new markings are Concorde G-BOAF and Boeing 757 G-BIKO.**
*British Airways*

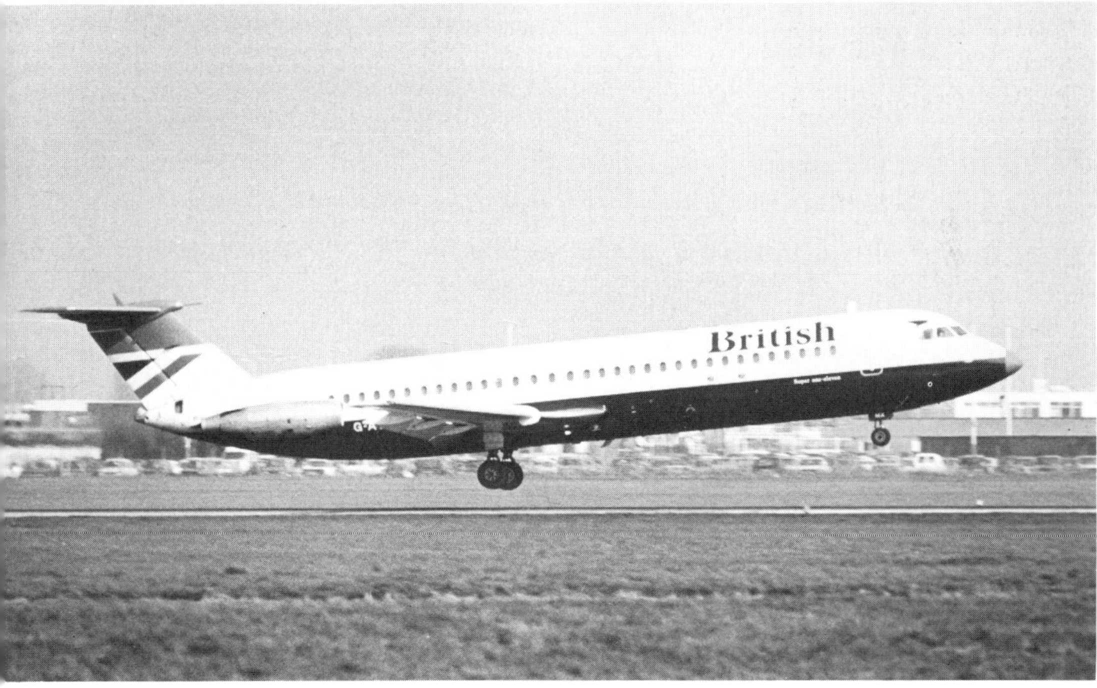


***Top:***
**British Airways Tristar taking off.** *British Airways*

***Above:***
**A British Airways BAC One-Eleven 500, designated the Super One-Eleven.** *British Airways*

airport, serving other airlines. The Highlands Division, serving the highlands and islands of Scotland with BAe 748s, is now in its third year of operation, and is beginning to show small but significant profits.

European services are split into North Europe and South Europe market centres. The former incorporates the airline's internal German services which link Berlin to seven points in West Germany. Flown with Boeing 737s, the services carry over one million passengers per year. West Berlin's Tegel Airport is British Airways' third busiest station anywhere in the world in terms of aircraft movements. South Europe comprises Portugal and all the Mediterranean countries served by British Airways with the exception of France. This market is largely composed of leisure travel, but also includes major business centres such as Madrid, Milan and Athens. North American routes include Concorde services to New York, Washington and Miami, and subsonic services to other points in the United States, Canada and the Caribbean. The airline's Caribbean routes are linked with new South American destinations, taken over from British Caledonian, from April 1985. Principal African destinations served by British Airways are Nairobi, Cairo and Johannesburg. Routes also extend eastwards to the Middle East, the Indian subcontinent, the Far East, and to Australia and New Zealand, giving a total unduplicated route network of over 500,000km.

British Airways carried nearly 16 million passengers in 1984 on scheduled services.

**Business Centres**

The Charter business centre serves three main functions which include the management of British Airtours Ltd, the chartering of aircraft from the mainline fleet, and the provision of handling services at Gatwick Airport for British Airways, British Airtours and for other airlines under contract. In addition to an extensive European programme, British Airtours also operates to the USA, Canada and the Caribbean, with a fleet of Boeing 737s and Lockheed TriStars.

The Cargo business centre handles a total volume of over 25,000 tonnes a year even though British Airways no longer has an all-cargo fleet. The airline has two major cargo centres at Heathrow and Manchester Airports.

Tours are combined collectively under the name of British Airways Holidays which includes the brand names of Sovereign, Enterprise, Flair and Martin Rooks. British Airways also owns a controlling interest in Speedbird Holidays.

**Aircraft Fleet**

The flagship of the British Airways fleet is, of course, Concorde. The aircraft are used on scheduled services to the United States as well as on increasingly popular charter flights (especially to the Bay of Biscay), and recorded a healthy profit during 1984. The last Concorde, grounded for three years, has now been put back into service, and all seven aircraft are programmed for a complete interior refit.

An outstanding success both in terms of passenger and crew appeal and operating costs has been the Boeing 757, first introduced in February 1983. Twenty-one of the type are now in service between Heathrow and a number of continental destinations, and they also link London with Aberdeen, Edinburgh, Glasgow and Manchester.

Other Boeing aircraft in the fleet include 28 747s, made up of 12 Rolls-Royce powered 747-236Bs and 16 747-136s with Pratt & Whitney engines, with a Combi or all-cargo 747 expected to be added shortly. Delivery of the last batch of 16 Boeing 737-200 twin-jets, ordered in autumn 1983, was completed in the spring of 1985, bringing the total in service to 44 aircraft. The 737s will replace the balance of the Trident fleet which cannot meet the new noise regulations which come into force on 31 December 1985. British Airways also holds options on another 15 737-200s.

Three new 748-2Bs were added late in 1984, and the two others owned by British Airways will be brought up to the same standard. Seventeen Lockheed TriStars and 26 BAC One-Elevens complete the fixed-wing fleet, totalling 148 aircraft. British Airways Helicopters' fleet consists of 31 helicopters including four Boeing-Vertol BV234 Chinooks, three Aerospatiale AS.332L Super Pumas, 17 Sikorsky S-61Ns, four Sikorsky S-76s and three Westland 30s.

TO FLY TO SERVE

# British Independent Airlines – Coming into their own

The year 1984 was dominated by the controversy surrounding the CAA review of UK civil aviation, the battle for routes between British Airways and British Caledonian and the rather muted contribution to the debate by the other independents, and the slippage yet again of the privatisation of British Airways, this time as a result of the law suits brought against BA and other airlines in the US courts by Sir Freddie Laker.

Historically, Britain's independent airlines have always had a rough deal, and the only airline to move into 1985 with any degree of satisfaction has been British Caledonian, whose route swap with British Airways must be regarded to have worked out in its favour. For the moment, the other private airlines have been bought off with promises of money (little though this will be) to develop the provincial links with Britain's European neighbours. Whilst the reasoning behind it is faulty and cannot be fairly justified, the privatisation of BA, now expected in this financial year, could well bring in its wake just enough loosening of state control all round to the direct benefit of Britain's private airlines.

## De-regulation?

Lower air fares on most domestic routes are on the way under new proposals to free airlines from the present strict CAA control over UK air routes and fares. The proposals first outlined in the CAA review published last summer will embrace all domestic routes except those to and from the Channel Islands and the Isle of Man. The two-year experiment came into effect on 1 April 1985. Airlines will be permitted to replace the existing system of making formal applications by simply giving 10 days' notice of any change in fares, but the CAA will retain powers to block any proposals which it considers predatory, or an abuse of monopoly powers.

These proposals would bring Britain much closer to US-style de-regulation, and the government is hoping that Britain's lead will encourage other EEC nations to follow suit. Taking the American experience as a guide, it would seem likely that fierce competition between Britain's biggest airlines will force down prices on the major routes between London and the provincial cities, to the detriment of inter-regional links, which make up about one-third of UK air routes. It is to be hoped that government would then step in to ensure that essential services are maintained in vital areas. This expected free-for-all may also initially encourage a proliferation of new cut-price airlines.

## British Caledonian Airways

Strong and dynamic, British Caledonian has come a long way in 25 years from a one-aircraft passenger and cargo charter operator in 1961, to Britain's major 'second force' international airline, currently employing 6,750 people. BCal's position has been further strengthened with the acquisition of the prime Saudi Arabian routes from British Airways in exchange for its non-profitable Latin American services, as a result of the much-publicised tussle with the national airline and the government over the CAA review on UK air policy. With a new service over the prestigious London-New York route inaugurated on 1 May 1985, the airline's international network now incorporates New York, Atlanta, Houston, Dallas/Fort Worth and Los Angeles in the United States; Dhahran and Jeddah in Saudi Arabia; and Hong Kong, operated via Dubai. In addition, British flag services to West, Central and North Africa serve the cities of Tunis, Tripoli, Banjul, Freetown, Monrovia, Abidjan, Accra, Kano, Lagos, Douala, Libreville and Lusaka. Domestic and European services also radiate from Gatwick to Amsterdam, Brussels, Frankfurt, Paris, Geneva and Genoa, and to Jersey, Manchester, Glasgow and Edinburgh. The airline is also associated with the British Airports Authority in the operation of the Gatwick-Heathrow helicopter airlink, and with seven commuter airlines in British Caledonian Commuter Services.

Passengers carried on all services during 1984 totalled 2.1 million, and extensive cargo operations accounted for 59,000 tonnes of freight and mail. BCal's fleet is made up of one Boeing 747-200B, three A310 Airbus, eight McDonnell Douglas DC-10-30s and 13 BAC One-Eleven 500s. The airline has initiated a £12million improvement programme for its One-Elevens which will include the fitting of hush-kits and the installation of automatic landing equipment.

***Above:***
**The two American-built wide-body jets in the B.Cal fleet include Boeing 747-200B G-BJXN and McDonnell Douglas DC-10-30 G-BFGI.**
*British Caledonian*

On order are seven A320 Airbus for delivery in 1987, and more DC-10-30s or Boeing 747s are likely to be acquired in the near future for its extended network.

British Caledonian is a wholly-owned subsidiary of Caledonian Aviation Group plc, which also incorporates British Caledonian Aircraft Trading, British Caledonian Travel Holdings (including Blue Sky Holidays), Caledonian Hotel Management, Caledonian Airmotive and **British Caledonian Helicopters**. Shareholders include Industrial & Commercial Finance Corporation, Royal Bank of Scotland, Bland Payne, Commercial Union, Sun Alliance and the Automobile Association. BCal has a minority holding in Gambia Airways for whom it provides management facilities. **British Caledonian Charter**, which uses two DC-10-10 aircraft of its own, is owned jointly with the Rank Organisation.

### The Trans-Atlantic Connection

With the demise of Laker Airways in February 1982, the trans-Atlantic routes were left once again in the hands of British Airways and BCal, neither airline offering a really cheap service. In 1983, following the introduction of a low-fare, no-frills service to London by the American airline, People Express, two airlines were formed in the UK to fill the void, **British Emerald Airways** and **British Atlantic Airways**. The former had hoped to fly from Belfast to New York, Miami and Los Angeles with Boeing 707s, and the latter, formed by Randolph Fields, intended to offer a luxury one-class service between Gatwick and Newark with ex-Laker DC-10s. British Emerald Airways still has not started, and the CAA also refused British Atlantic a licence. Fields appealed against the decision and persuaded the colourful President of the giant Virgin Group, Richard Branson, to invest a seven-figure sum in the airline. To Richard Branson, who originally introduced the concept of low-price records to the UK, owning a low-fare airline provided a new but not dissimilar challenge. In typical showbiz style Richard Branson and Randolph Fields announced, in April 1984, the birth of **Virgin Atlantic Airways**. Fields has since left pursuing more ambitious plans by founding **Highland Express** at Prestwick, to offer a similar service to New York and Toronto with a fleet of Lockheed TriStars beginning in 1986. Originally refused a licence, this new airline has been granted a re-hearing. Both British Airways and **British Midland Airways** have objected, the latter to safeguard its planned services from Manchester and Glasgow to New York due to have started in April 1985.

Virgin Atlantic obtained its licence from the CAA and began operations with a Boeing 747 on 22 June 1984. The service operates daily from 1 April and throughout the summer, with fewer flights in winter. Three classes of service are currently offered: Upper Class, Space Class, and Economy Class. Space Class is a new low-fare concept for the business traveller, guaranteeing an empty seat next to the passenger. So far the service has been a success with load factors averaging out at around 90%

In order to boost its winter traffic across the Atlantic, the airline has introduced a Maastricht-Gatwick feeder service which it hopes will generate at least 10% of the Gatwick-Newark traffic.

### Euro-Link

Since the very first London-Paris service by Aircraft Transport & Travel in 1919, airlines have always viewed the Paris connection as their prime objective. As many as four airlines have at one time competed for traffic between London and the French capital, and it remains to this day the UK's busiest connection. Both **British Airways** and **British Caledonian** now provide frequent daily flights, seven days a week, from Heathrow and Gatwick respectively, and Air UK has joined the fray with a service from nearby Stansted. BA and BCal also fly many other services from London to most European capitals.

Direct links are now also provided from a number of provincial cities. British Midland Airways serves Paris during weekdays from East Midlands, Belfast and Glasgow, using Fokker Friendships; Air UK flies direct from Edinburgh, Leeds/Bradford, Newcastle and Southampton; and Jersey European Airways has added a service from Southend.

The total number of passengers carried between the UK and Paris in 1984 was 2,144,149, of which 2,132,012 flew on scheduled services.

It is in fact the provincial cities which over the past years have added the missing links by offering direct flights to many other European cities to the enormous benefit of business and trade. **British Midland Airways**, one of Britain's longest-established airlines having been founded in 1938 as Derby Aviation, has in addition to the Paris routes, services from Birmingham to Brussels, and from East Midlands to Amsterdam. BMA is now one of the largest privately-owned airlines, employing 1,400 people, and carrying over two million passengers annually. Although based at East Midlands Airport, the airline operates principally from London Heathrow, where it is the second busiest carrier after British Airways with services to

Virgin
G-VIRG

*Left:*
**Virgin Atlantic uses this Boeing 747-200B, G-VIRG, on low-cost trans-Atlantic flights between Gatwick and Newark.** *John Cook*

*Top:*
**A line-up of Air UK Fokker Friendships at Schiphol Airport, Amsterdam.** *Air UK*

*Above:*
**Air UK also operates the Friendship's bigger brother, the F.28 Fellowship.** *Air UK*

Glasgow, Edinburgh, Belfast, Birmingham, East Midlands, Leeds/Bradford, Liverpool and Tees-side. Further north, **Aerotime** has obtained a licence to link Edinburgh with Copenhagen, Brussels and Frankfurt.

**Air UK**, formed through the merger of British Island Airways (since bought out and re-formed), BIA/Air West, Air Anglia and Air Wales, has an extensive network of services which link most major points in the UK and several points in Europe; these, in addition to Paris, include Amsterdam from Aberdeen, Edinburgh, Humberside, Newcastle, Norwich, Stansted and Southampton; Brussels from Stansted; Copenhagen from Aberdeen and Newcastle; Esbjerg from Heathrow and Humberside; and Bergen and Stavanger from Aberdeen. The airline employs some 850 people.

**Dan-Air**'s European services extend from Newcastle to Stravanger and Bergen; Tees-side to Amsterdam; Manchester to Zurich; and from London Gatwick to Berne and Zurich in Switzerland, and to Montpelier, Perpignan and Toulouse in southern France. Innsbruck will be added in December 1985. Dan-Air also flies between West Berlin, Amsterdam and Saa-brücken.

The Midlands are also served by **Birmingham Executive Airways** which commenced services in June 1983, and now links Birmingham with Copenhagen, Stockholm, Stuttgart, Geneva, Zurich and Milan, using BAe Jetstreams, and a Saab-Fairchild SF-340 aircraft configured in a special 27-seat business class layout. Recent newcomers on the European scene include **Brown Air Services** with several weekly flights between Leeds/Bradford and Frankfurt and Oslo, and **London European Airways**, flying between Luton and Amsterdam with a Vickers Viscount, and over the former Euroflite routes to Brussels from Luton and East Midlands with a Jetstream. A new Ipswich-based airline, **Suckling Aviation**, hoped to serve Amsterdam from spring 1985 with a Twin Otter, and **EEC Airways** has applied for routes from Stansted to Rotterdam, Hannover, Cologne and Luxembourg. **Guernsey Airlines** is planning a Guernsey-Paris route, and **British Air Ferries**, which operates a Gatwick-Rotterdam service, is looking to connect Coventry with Basle, Beauvais, Lille and Ostend.

### Flying to the Sun

Britain's holiday charter airlines are having to face up to even more difficult times in 1985 than they experienced last year. The weakness of the pound sterling against almost all currencies the countries most favoured by the British holiday-maker, will mean increased tour prices and reduced spending power in, for example, France, Spain, Portugal and Italy. The exceptions to this are the Balkan countries like Greece and Yugoslavia, but overall tour bookings are reported to be heavily down on the same period in 1984. Competition among charter airlines is also hotting up with a decisive thrust into the market being mounted by British Airtours, the charter subsidiary of British Airways. The first airline to suffer from this was **Air Europe**, when its parent company, the giant Intasun Group, struck a five-year flights deal in the summer of 1984, giving British Airtours a quarter of its business. Air Europe will continue to fly the major part of the Intasun tour programme, now amounting to some 40%. IT flights are also undertaken for a number of other prominent tour operators. The airline's staff and aircraft fleet for the 1985 season has been drastically reduced as a result of the above loss and general overcapacity in the market, and will comprise three 737-200s and two 757-200s. Scheduled services to Palma de Mallorca were in inaugurated in April 1985, and a Gatwick-Gibraltar service is expected to begin in November 1985.

**British Airtours** was set up in 1969 to fulfil British Airways' charter requirements for its own tour programmes marketed under the Sovereign, Enterprise, Flair, Martin Rooks and Speedbird Holidays labels, and these still command the largest shares of Airtours business with 30%. The Intasun contract takes up 20% of the carrier's capacity, and other substantial deals have been concluded with Thomson Holidays and Horizon. British Airtours operates out of Gatwick Airport with a fleet of Lockheed TriStars and Boeing 737s.

Horizon Travel of Birmingham also has its own airline, **Orion Airways**, formed on 28 November 1978. Orion Airways is based at East Midlands Airport where, apart from its administrative offices, it also maintains its own training and engineering facilities, employing a staff of 400. In addition to East Midlands traffic, the airline also has major tour programmes originating from most other provincial UK airports, carrying over 1,500,000 passengers a year for its parent company and other tour operators. The all-jet holiday airline began operations in March 1980 with a fleet of four brand-new Boeing 737-200 twin-jets. The first of five Boeing 737-300s on order entered service in February 1985. This newest version of Boeing's 737 family is powered by more fuel-efficient engines and seats 144 passengers in an all-charter layout. Orion was the first non-US customer to order this new technology aircraft.

Two companies which were very much involved in the development and boom years of

*Below:*
**Air Europe operates on all-Boeing fleet of 757 and 737 twin-jets on inclusive-tour and charter flights. 737-200, G-BMHG is one of three currently in service.** *Boeing*

holiday traffic are **Dan-Air Services**, based at Gatwick, and **Britannia Airways**, with headquarters at Luton. The former was founded on 21 May 1953 as a subsidiary of Davies & Newman, the London shipping brokers, from whom its name is derived. Dan-Air has developed extensive inclusive-tour flights which are now provided from 13 points in the UK and West Berlin for a number of leading British and German tour operators. The airline also operates contract services for oil companies, night mail flights, and passenger and cargo charters. A growing network of passenger schedules serves 28 cities in seven countries with a fleet of 50 aircraft.

Britannia Airways began operations on 5 May 1962 with Lockheed Constellations, and has been a wholly-owned subsidiary of the Thomson Organisation since 26 April 1965. The airline has since grown to become Britain's largest charter operator with a modern fleet of three Boeing 767s and 29 Boeing 737-200s, carrying 4.5 million passengers annually. Regular operations cover more than 50 destinations from all parts of Britain, and include all the principal holiday resort areas. About 50% of its business is made up of flights for Thomson, with the remainder for other major tour companies. A scheduled service has been added linking Manchester and Palma, with Malaga to follow.

In addition to IT and charter services, the airline is also involved in MoD trooping contracts between the UK and Germany and Gibraltar, and aircraft leasing, maintenance and training. A Boeing 737 is operated on behalf of GB Airways on the three times a week service between Gatwick and Gibraltar. Another Luton resident is **Monarch Airlines**, which undertakes inclusive-tour and worldwide charter services from all major points in the United Kingdom, with a fleet comprising four Rolls-Royce powered Boeing 757s and four Boeing 737-200s. Principal destinations are Europe and the Middle East. The airline, associated with the tour operator Cosmos, was formed on 5 June 1967 and began operations the following April with two Bristol Britannias. Maintenance and engineering services for Monarch and other airlines are provided by its associate company, Monarch Aircraft Engineering.

Inclusive-tour and charter flights, together with scheduled cargo services, are provided from Southend Airport by **British Air Ferries (BAF)** with a fleet of Vickers Viscounts and Handley Page Heralds. Now a subsidiary of the property company, Jadepoint Ltd, BAF was formed in January 1963 through the merger of Channel Airbridge and Silver City Airways, both of whom built their reputation with cross-Channel car and passenger services. BAF was known as British United Air Ferries until September 1967.

Lydd Airport in Kent is the home of **Janus Airways** which flies Heralds on IT operations to the continent. The airline is a subsidiary of the Midlands company, Hards Travel Service, and initiated services on 2 January 1983.

**British Island Airways**, based at Gatwick, utilises eight One-Elevens on holiday and ad-hoc charter flights to European and North African destinations. The airline also provides specialist services to the oil industry, aerial ambulance, and wet-lease facilities for other airlines.

Wales now has its own holiday airline, **Airways International Cymru**, formed by Red Dragon Travel. Two One-Elevens and a Boeing 737 twin-jet are used on IT flights from Cardiff-Wales Airport. Operations began on 1 February 1984.

**London Express Aviation** has applied for low-cost charter flights from Stansted to Singapore and Hong Kong beginning on 4 October 1985, but is likely to face stiff opposition from British Airways, British Caledonian and Cathay Pacific.

### Cargo Operations

It is a measure of the fluctuating fortunes of the air cargo business in the 1970s that few all-cargo airlines in Britain survived that particularly difficult decade. In stark contrast the 1980s have witnessed a sharp increase in traffic mainly due to the emergence of the small parcel express and courier market, reaching a peak in 1983, with a total of 350,000 tonnes being carried by the major passenger airlines, as well as a handful of specialist cargo companies.

The outlook for 1986 is considered to be favourable with an expected growth rate of 10%, though the industry will be faced with a fair amount of uncertainty brought about by the new noise regulations, and the consequent elimination of night-flying, traditionally the best time for shippers. Today's highly competitive environment also highlights an increasing volume of time-sensitive items, and it is the experienced and best equipped charter airlines which stand to benefit most.

Belly cargo space in passenger aircraft is being offered by all major airlines such as British Airways, British Caledonian, Dan-Air Services, British Midland Airways and Air UK, making up a sizeable proportion of the total UK cargo traffic. B.Cal and Dan-Air in particular are strongly promoting their cargo-carrying capability, and British Airways is considering re-entry into the all-cargo business with a Boeing 747 Combi.

Britain's oldest all-cargo airline is **Tradewinds Airways**, now a subsidiary of the Lonrho

***Top:***
**Orion Airways' first Boeing 737-300, G-BLKB, was delivered early in 1985 in time for the summer season.** *Orion*

***Above:***
**Britannia Airways, Britain's biggest holiday airline, is the only UK operator of the Boeing 767. It has three in service including G-BKPW.**
*Boeing via David Dorrell*

*Below:*
**Dan-Air is the UK's only operator of the Boeing 727 tri-jet. G-BHNF is one of 10 in the fleet.** *Dan-Air*

organisation, which made its first commercial flight on 2 April 1969 with a charter to Brisbane, Australia. A network of regular split-charter services is now operated to destinations in East and West Africa including Khartoum, Mogadishu, Lagos, Kano, Port Harcourt and Lusaka, together with ad-hoc cargo flights worldwide. The airline also specialises in the transportation of horses.

In February 1983 Tradewinds was granted permission by the CAA to operate scheduled cargo services to a total of 21 destinations in Africa, North America, Canada and the Middle East. Of these it currently serves Chicago, Toronto and Khartoum with others planned to follow in the near future. Fleet consists of Boeing 707s with a payload capacity of 40 tonnes.

Also based at Gatwick is **Anglo Cargo**, which offers long-haul charter services with a single Boeing 707. The airline began operations in January 1984, initially concentrating on the Middle East and African markets.

Resident at East Midlands Airport, Castle Donington, is **Air Bridge Carriers**, which not only survived the 1970s, but has grown into a major cargo carrier with its own modern maintenance facility, and currently employs 100 people. Air Bridge was formed in November 1972 by Field Aviation with a fleet of four Armstrong Whitworth Argosies, initially to operate between East Midlands and the Channel Islands. Operations have since been expanded considerably and, with a mainline fleet of Vickers Merchantman aircraft capable of carrying a payload of 18,500kg, the airline now provides world-wide ad-hoc cargo flights with the transportation of race horses and other livestock being a speciality. Contract work comprises nightly newspaper flights between Luton and Glasgow, and express parcel services for overnight courier companies.

East Midlands Airport is also the hub for the air-road network of **Elan International** which specialises exclusively in the overnight delivery of consignments, door-to-door, between the UK, Ireland, Belgium, Holland and Luxembourg. In May/June 1984 the company acquired its own fleet of aircraft in the shape of a Handley Page HPR.7 Herald with a payload of 5,000kg, and an Armstrong-Whitworth AW.650 Argosy with an 11,000kg payload. Also available is a Vickers V.953C Merchantman which has a payload of 18,500kg. All aircraft are operated by Air Bridge Carriers. At present the Dart Herald operates between the Benelux countries and the UK, with the Argosy serving the Republic of Ireland. Freight between Belfast and the UK mainland is flown on the Merchantman.

**Skyguard** uses a Herald on an overnight Eurosprint parcel service between Birmingham and Brussels.

Guernsey's States Airport and Bournemouth's Hurn Airport are home to **Channel Express Air Services**. The airline was formed in January 1978 as Express Air Services and, after selling its passenger operations to Jersey European Airways, concentrated exclusively on cargo work. This currently includes frequent regular weekly round trips between Bournemouth and Channel Islands, with special expertise in handling perishable goods, and a scheduled night mail service over the Bournemouth-Bristol-Liverpool route. Dart Heralds are available for charter or lease in passenger, freight or combi configuration, and a five-passenger Partenavia is operated on executive charter throughout Europe.

**Atlantic Air Transport** is the name of the cargo division of the Jersey-based company General Aviation Services. From a base at Stansted the airline utilises a fleet of Douglas DC-3s and a single Bristol Freighter on ad-hoc cargo charter flights to Europe, North Africa and the Middle East. During the summer only, two DC-3s are converted to passenger use, and another for survey work, in which configurations they are operated under the name of **Air Atlantique**.

Atlantic Air Transport's first commercial flight was made on 16 May 1984 from Gatwick to Dublin, carrying a racehorse entered for the Irish 2000 Guineas Classic.

**British Air Ferries** flies a scheduled cargo service between Jersey, Dusseldorf and Belfast, together with regular flights from Southend to Cologne and Saarbrücken in Germany, and to Basle in Switzerland. Contract work is also undertaken on behalf of Air UK and TNT Courier. A large fleet of four-engined Vickers Viscounts is used.

The Luton-based airline **London European Airways** is also making a bid for cargo business on its new Viscount service to Amsterdam.

### Too Large to Carry?

The answer to this question would at one time have been emphatically yes. At least until **HeavyLift Cargo Airlines** came on to the scene in 1978, at the same time as the RAF withdrew from service its 10 Belfast freighters. Capt P. J. McGoldrick, now Deputy Chairman of HeavyLift, realised the potential of the massive cargo capacity of the aircraft, and with single-minded determination set out to put his ideas into practice. It proved a long and difficult, but ultimately successful, project, with the Belfast receiving its CofA in 1980. Most of the technical conversion and certification work was carried out

***Top:***
**Monarch Airlines operates four Boeing 757-200s, including G-MONB pictured, as well as G-MONC, G-MOND and G-MONE.** *Leeds/Bradford Airport*

***Above:***
**Two Gatwick-based cargo airlines, both utilising Boeing 707s, are Tradewinds Airways and Anglo Cargo.** *Tradewinds/Anglo Cargo*

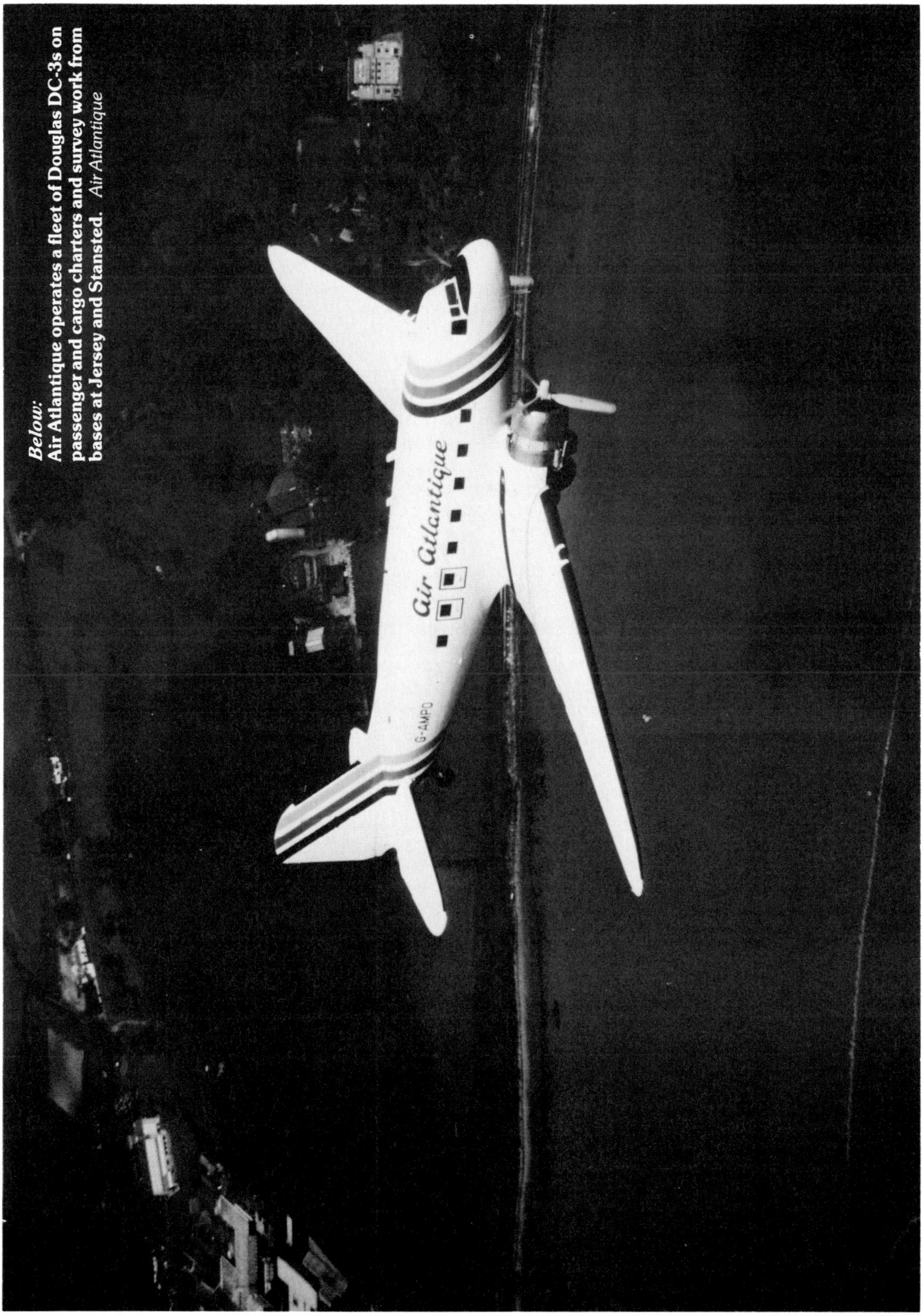


*Below:*
**Air Atlantique operates a fleet of Douglas DC-3s on passenger and cargo charters and survey work from bases at Jersey and Stansted.** *Air Atlantique*

by Marshall's of Cambridge, and the Belfast entered service in March that same year, recording almost immediate success. HeavyLift now owns five aircraft, of which it keeps three in regular service, together with a Canadair CL-44-0 Guppy.

The Belfast has a virtally unrestricted volume hold of 350cu m and a payload capacity of 36,500kg, which enables it to carry complete air frames, helicopters and engines with the minimal disassembly. This capacity makes it particularly attractive to aerospace manufacturers, and about 30% of the airline's current business is derived from work in this area. The oil industry also provides much of the Belfast's freight. For the future, the regular transportation of satellites and rocket parts is very much under consideration.

HeavyLift, owned by the Cunard Steamship Co (two thirds) and Eurolatin Aviation, employs around 70 people at its head office at Stansted Airport and an associated maintenance organisation at Southend Airport, also in Essex.

Main markets for the airline's operations are in Europe, the Americas and the Far East.

**Datapost**

Datapost is the Royal Mail courier service for nationwide next day delivery of mail and small packages which, apart from the usual rail and road distribution service, also includes a unique air network of specially chartered aircraft.

The hub of the Datapost network is at Luton Airport from where aircraft leave two hours either side of midnight, Mondays to Fridays, to Edinburgh, Glasgow, Manchester, East Midlands, Belfast, Dublin and Bristol. There is also a Luton-Rotterdam link. The network is being serviced by several contractors, each operating a specific route. **Air Ecosse** operates the services to Edinburgh, Aberdeen and Dublin with the Shorts 360 which has a payload of 3,500kg and the Embraer Bandeirante with a capacity of 1,500kg. **Manx Airlines** flies the Luton-Manchester service, and **Loganair**, the Luton-Glasgow route, both using the Shorts 360. The 1,600kg Twin Otter is scheduled by **Spacegrand** on the Luton-East Midlands-Belfast round trip, whilst the Elstree-based company **National Airways** uses a similar capacity Trislander between Luton and Bristol. The Luton-Rotterdam link is currently flown by **Aviation West** with a Piper Navajo, and the latest addition to the network is a City-Heathrow connection by **Air Hanson** with a 375kg payload Bell JetRanger.

**Britain's Growing Commuters**

Britain's commuter operations, for many years the forgotten sector of UK civil aviation, are at last beginning to come into their own. A look at the services now available in the United Kingdom reveals a dense system of routes linking provincial centres to London and across country to each other. Encouragement is provided by the fact that even the smaller communities which hitherto have been unable to sustain traffic of sufficiently economic proportions have now also been incorporated into the countrywide network.

This growth over the last five years can be directly attributed to the shedding by British Airways of some of its lesser and historically unprofitable routes combined with the advent of smaller and purpose-built aircraft such as the Bandeirante, Shorts 330, Swearingen Metro and more recently the Shorts 360, British Aerospace Jetstream 31 and SAAB-Fairchild SF-340. The latest types to enter service, although not yet in the UK, such as the ATR-42 and Embraer Brasilia among others, will offer even more flexibility at a higher maximum capacity than the 30-seater available until recently. The third factor to bring about this more buoyant picture is the existence of several co-operative ventures between commuters and larger airlines to the mutual benefit of both.

Small commuter airlines have always been faced with an expensive UK aviation infrastructure which is geared mainly to operators of large aircraft. As yet the authorities have not quite recognised the special, and, in many instances considerably less demanding, needs of the commuter airlines, whose cost burdens are thus out of proportion to their requirements for services. The commuter airlines' unique position and importance has been acknowledged by the bigger airlines such as B.Cal, Dan-Air and British Midland, but for a continuing and sustained development of regional air services in the United Kingdom a fundamental re-assessment of the country's civil aviation structure will be essential.

**British Caledonian Commuter**

B.Cal Commuter is a joint operation of British Caledonian and seven other airlines providing a comprehensive and integrated network of internal air services. The present route system combines the operations of **B.Cal, Aerotime, Air Ecosse, British Air Ferries, Brymon Airways, Connectair, Guernsey Airlines** and **Jersey European Airways**, offering almost blanket coverage of the United Kingdom. Together with B.Cal's trunk route operation linking Gatwick with Glasgow, Manchester, Edinburgh and Jersey using 99-seat BAC One-Elevens, the commuter system serves 10 points in England, Wales, Scotland and the

***Above:***
**Air Eçosse uses this Shorts 360, G-RMSS, on Datapost flights from Luton.** *Shorts via Mike Hooks*

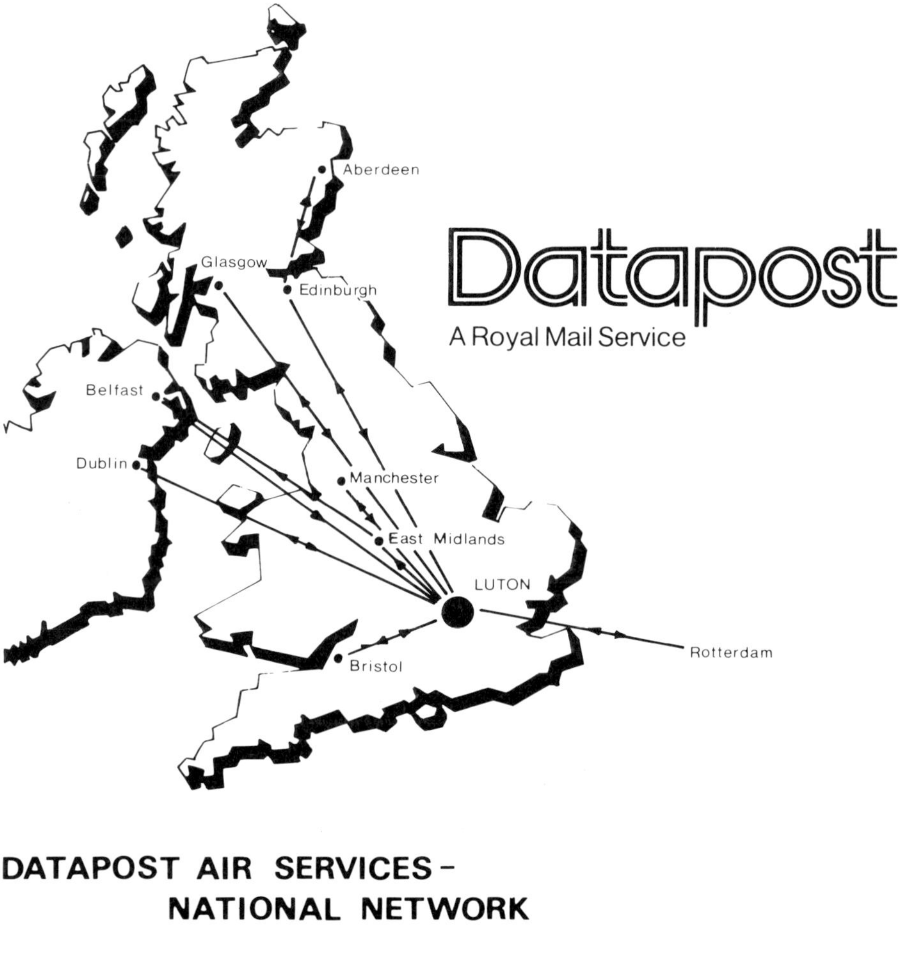


**DATAPOST AIR SERVICES – NATIONAL NETWORK**

# BRITISH CALEDONIAN COMMUTER NETWORK

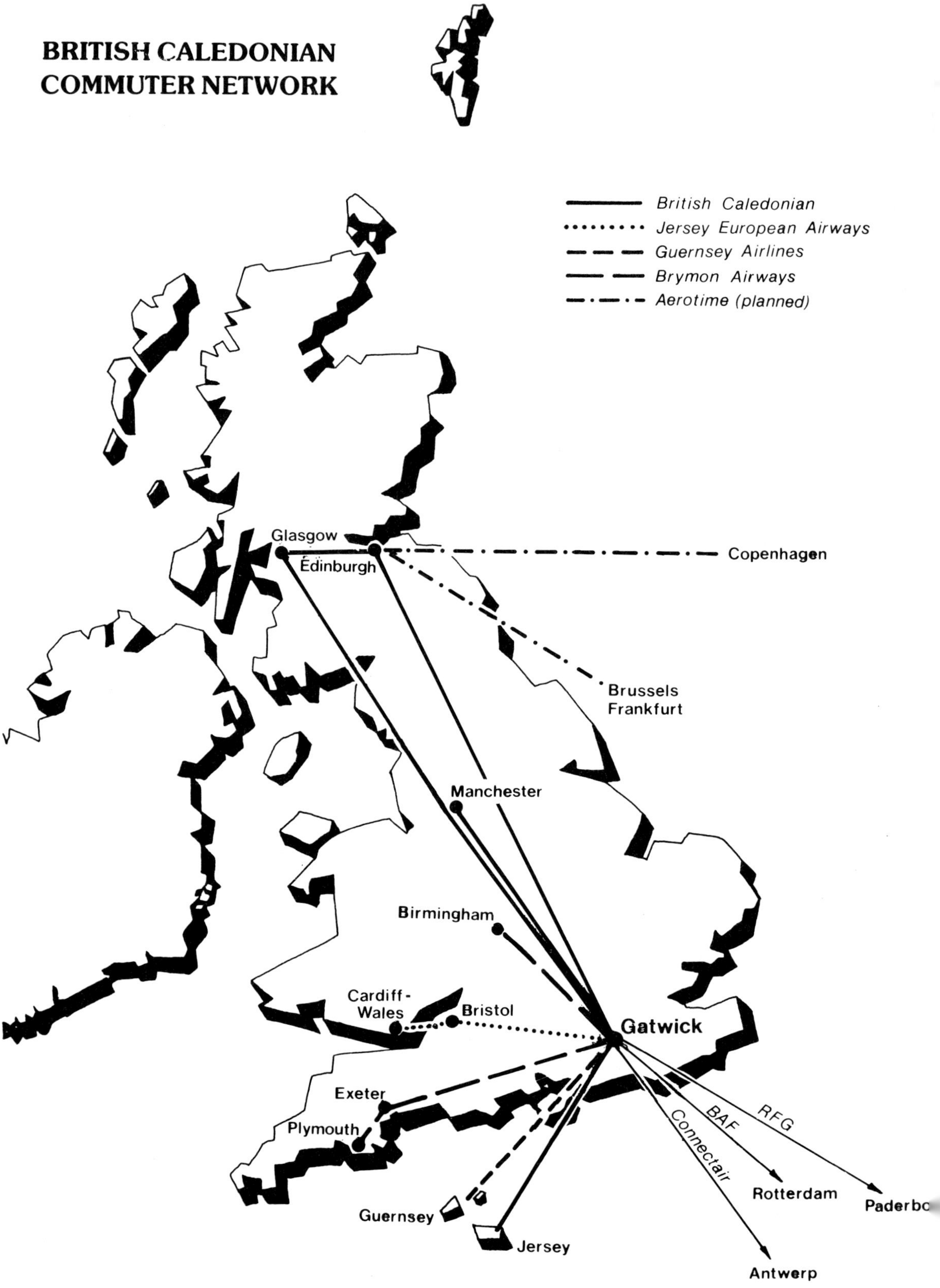

Channel Islands direct from Gatwick. Also included are international connections to Antwerp in Belgium and Rotterdam in the Netherlands. **R.F.G.** of Dortmund is also a member, and now connects Gatwick with Paderborn in West Germany.

The B.Cal Commuter network is centred on the airline's London Gatwick hub and all services are timed to provide easy and convenient connections not only with its international and intercontinental services by also with those of other long-haul airlines. The current commuter fleet comprises aircraft ranging from 18-seat to 99-seat capacity and includes the Shorts 330, Embraer Bandeirante, de Havilland Canada Twin Otter, Swearingen Metroliner, and BAC One-Elevens and Vickers Viscounts.

As part of this multi-airline agreement, all these services are featured in B.Cal's worldwide reservations system and timetable. Although a contract is drawn up between the participating carriers, the arrangement is flexible in that B.Cal has no financial interests and all routes are those of the commuter airlines themselves, and not replacement routes (as in the USA) operated on its behalf. The one provision is that no route must be operated in direct competition. In exchange for the obvious increase in interline traffic revenue, B.Cal makes available its substantial resources to the smaller airlines and assists with and advises on ground handling at Gatwick: advertising, sales and marketing; planning, scheduling and licensing of routes; revenue accounting; insurance; staff training; tariff applications to IATA; communications; cargo services; and buying, selling and leasing of aircraft through British Caledonian Aircraft Trading. B.Cal also provides emergency engineering services at Gatwick but is not contracted to carry out major engineering work. Some of these services are provided at cost, the real benefit to B.Cal accruing from increased use of Gatwick Airport and consequent growth of interline passenger traffic.

The first commuter airline to be associated with B.Cal in this new scheme was the Liverpool-based **Genair** which from 2 November 1982 began operating its aircraft in the new British Caledonian Commuter Services livery on its routes linking Gatwick with Humberside, Leeds/Bradford, Liverpool, Norwich and Teesside. Genair has unfortunately since fallen by the wayside.

**Dan-Air's** Link City services operated on a similar scale, and comprised services from Bournemouth to Birmingham, Cardiff, Manchester, Newcastle and Glasgow, and between Glasgow, Leeds/Bradford, Cardiff and Bristol. All these are now flown by **Metropolitan Airways** in its own right using Shorts 330s. **Air UK Commuter** obtained the Genair licences on the Heathrow-Norwich-Humberside and Humberside-Esbjerg routes in 1984, on which it uses Shorts 330s and 360s.

Through its purchase of Loganair and the formation of Manx Airlines jointly with Air UK, **British Midland** has also staked its claim to a slice of the commuter and regional market. In addition it operates Shorts 360s in its own colours between London and East Midlands and Birmingham.

**Loganair**, Scotland's principal airline, operates a comprehensive network centred on Glasgow and Edinburgh, serving Scotland, the North of England, Northern Ireland and the Isle of Man. Extensive inter-island services are flown in the Orkneys and Shetlands, and there is a Western Isle route. Additionally, Loganair operates the Scottish Air Ambulance service. At the other end of the UK,**Aurigny Air Services** provides inter-island services between the three main Channel Islands of Jersey, Guernsey and Alderney, as well as flights to mainland Britain and northern France. **Manx Airlines** flies frequent connections from the Isle of Man to London Heathrow and several provincial cities, using Shorts aircraft.

**Eastern Provincial Airlines** has applied for scheduled routes from Humberside to London, Norwich, Tees-side and Glasgow, and across the North Sea to Esbjerg in Denmark. Among the most recent additions are services by **Air Camelot** between Bristol and Exeter, Bournemouth, Alderney and Cherbourg with Trislanders, and by **Welsh Airways** between Aberport, Cardiff, Haverfordwest, Shobdon, Llanbedr and Swansea. Welsh Airways also hopes to link Wales with London, Aberdeen, and with Waterford in Ireland. **Air Wight** of Hurn has applied for scheduled licences between Bembridge and Bournemouth and Southampton. **Air Eçosse, Ace Aviation, Peregrine Air Services** and Gatwick-based **Euroair**, have applied for a number of scheduled services from Aberdeen to a range of European destinations. Although not a UK airline, **Aer Lingus Commuter** (formed in March 1984) connects Bristol, Edinburgh, Leeds/Bradford and Liverpool to Dublin using Shorts 330s and 360s.

For the small commuter airlines, close co-operation with the major airlines has brought considerable benefits. There is no doubt though that a well organised regional commuter system benefits not only the commuter airlines, but also the larger interline carriers and, most important of all, the travelling public.

# DOMESTIC NETWORK

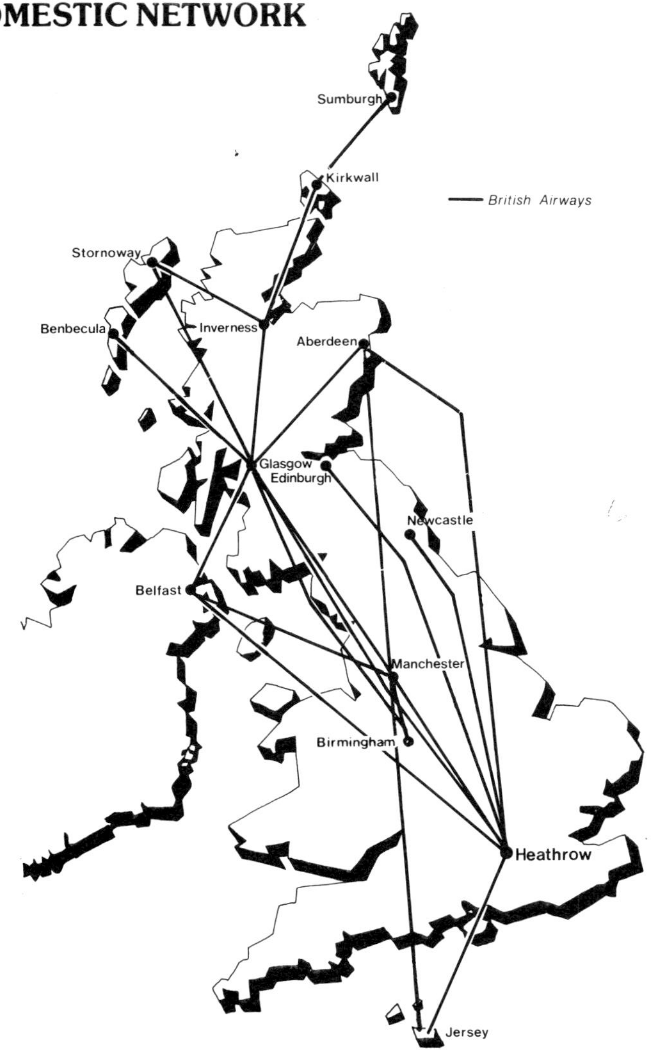


# DOMESTIC NETWORK

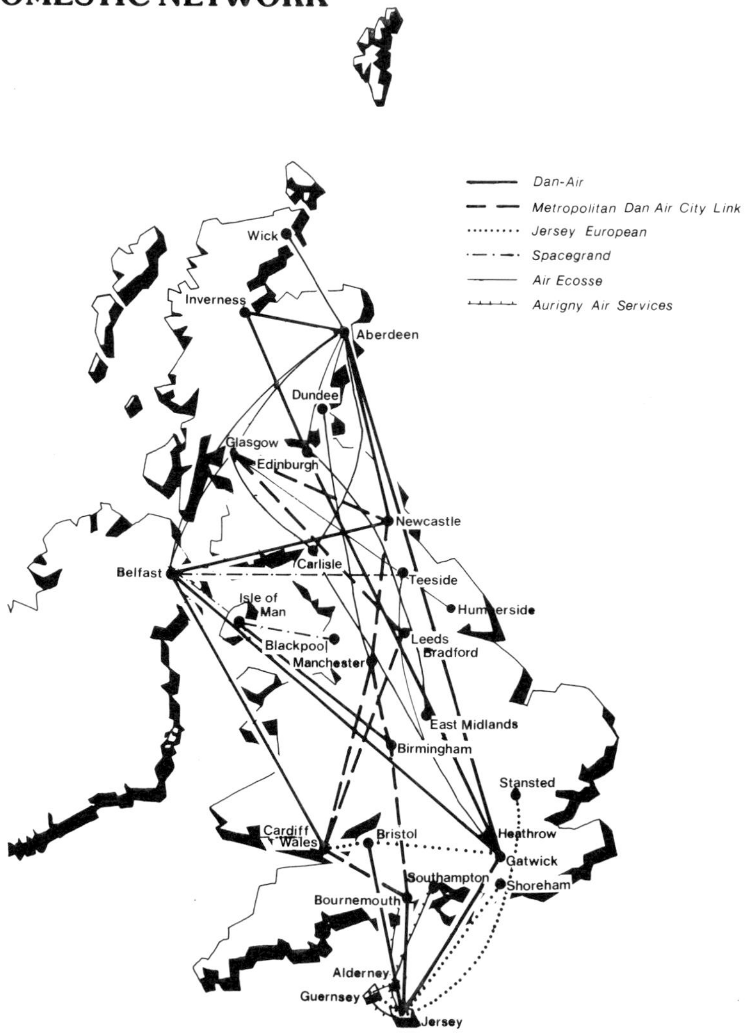

*Below:*
**A fine aerial study of Manx Airlines' Shorts 360, appropriately registered G-ISLE.** *Manx Airlines*

The recent upswing of the commuter industry has resulted in the formation of two associations devoted to furthering the interests of the commuter airlines. The first to be established was the **British Commuter Airlines Association** on 25 April 1984. Present membership includes Air Eçosse, Aurigny Air Services, Brymon Airways, Jersey European Airways, Loganair, Metropolitan Airways and Spacegrand Aviation. The driving force behind the association was David Beety, Chairman of Metropolitan Airways, who recognised that only as a strong united body can the commuters hope to achieve desired recognition and improvements. The BCAA's aims are to promote interest and understanding in the operation of commuter airlines, and in particular to co-ordinate matters of common concern among operators whose operations are geographically diverse. Among the areas the BCAA hopes to improve are product support, CAA requirements and route navigation, and other infrastructure changes.

The **Gatwick Regional Airlines Association** was formed in 1984 at the instigation of Mike Kay of British Air Ferries. It is basically a grouping of commuter airlines operating into Gatwick Airport, and presently incorporates Brymon Airways, Connectair, Jersey European Airways, Guernsey Airlines and British Caledonian. The association meets at three-monthly intervals to discuss all matters concerning operations at Gatwick, and high on the list of priorities are better terminal facilities. At present they are housed well away from the main terminal, in the general aviation area dubbed with some justification the 'playpen'. With the second terminal due for completion by 1987, talks are being held with the BAA with the aim of obtaining suitable accommodation in this new building. GRAA is also well aware of the need for rationalisation and a degree of standardisation of members' aircraft, and hopes that jointly its members will be able to negotiate better deals when buying equipment. To this end each airline is represented by a member from both the commercial and engineering sides.

*ADDRESSES*

**Aerotime Ltd** (1979)
Head Office: General Aviation Terminal, Edinburgh Airport, Edinburgh, Scotland EH12 9DN
Fleet: One Beechcraft 65-C90 King Air. On order: Two SAAB-Fairchild SF-340As.

**Air Bridge Carriers Ltd (AG)** (1972)
Head Office: Cargo Terminal 2, East Midlands Airport, Derby DE7 2SA
Fleet: Five Vickers V953C Merchantmen, two Armstrong Whitworth AW.650 Argosies.

**Air Camelot** (1978)
Head Office: Terminal Building, Bristol Airport, Bristol, Avon BS19 3DY
Fleet: Two Britten-Norman BN-2A Mk III Trislanders.

**Air Eçosse Ltd (SM)** (1977)
Head Office: Dyce Airport, Aberdeen, Scotland AB2 0PB
Fleet: Three Embraer EMB-110 Bandeirantes, two Shorts 360s, four Shorts 330s.

**Air Europe (AE)** (1978)
Head Office: Europe House, East Park, Crawley, Sussex RH10 6AS
Fleet: Two Boeing 757-236s, three Boeing 737-2S3s.

**Air Luton Ltd** (1984)
Head Office: Luton International Airport, Luton, Bedfordshire LU2 9PA
Fleet: Two Douglas DC-3s

**Air UK Ltd (UK)** (1980)
Head Office: Cross Keys House, Haslett Avenue, Crawley, Sussex RH10 1HS
Fleet: Four Fokker F.28 Fellowships, 14 Fokker F.27 Friendships, four Shorts 360s, four Shorts 330s.

**Airways International Cymru Ltd (AK)** (1984)
Head Office: 10 Church Street, Cardiff CF1 3QD, Wales
Fleet: Two BAC One-Eleven 304AXs, one Boeing 737-204.

**Anglo Airlines Ltd (ML)** (1983)
Head Office: Unit 15, Fleming Way, Crawley, West Sussex RH10 2NX
Fleet: One Boeing 707-338C.

**Atlantic Air Transport/Air Atlantique (DG)** (1977)
Head Office: Hangar 31, Stansted-London Airport, Bishops Stortford, Essex
Fleet: Six Douglas DC-3s, one Bristol 170 Mk 31 Freighter.

**Aurigny Air Services Ltd (GR)** (1968)
Head Office: Alderney, Channel Islands
Fleet: Nine Britten-Norman BN-2A Mk III Trislanders, two Britten-Norman BN-2A Islanders.

**Birmingham Executive Airways Ltd (EB)** (1983)
Head Office: Forward House, Birmingham International Airport, Coventry Road, Birmingham B26 3QJ
Fleet: Three BAe Jetstream 31s, one Saab-Fairchild SF-340A.

**Britannia Airways Ltd (BY)** (1961)
Head Office: Luton Airport, Luton, Bedfordshire LU2 9ND
Fleet: Four Boeing 767-204s, 26 Boeing 737-204s.

***Top:***
**Air Camelot/Aviation West Trislander G-BEVV serves several cities in the southwest, as well as Alderney in the Channel Islands, from Bristol.**
*Aviation West*

***Above:***
**Aurigny Trislander G-BEPI photographed at Jersey. This Alderney-based airline has been serving the Channel Islands since 1968.**
*K. Siskell via Aviation Data Centre*

**British Air Ferries Ltd (VF)** (1963)
Head Office: Municipal Airport, Southend-on-Sea, Essex SS2 6YL
Fleet: Nineteen Vickers V.800 Viscounts, two Handley Page HPR.7 Herald 214s.

**British Airtours (KT)** (1969)
Head Office: Airtours House, Gatwick Airport, West Sussex RH6 0LF
Fleet: Three Lockheed L.1011 TriStars, 13 Boeing 737-200s.

**British Airways (BA)** (1924)
Head Office: Speedbird House, London Airport Heathrow, Hounslow, Middlesex
Fleet: Seven BAe/Aerospatiale Concorde 102s, 21 Boeing 757-236s, 28 Boeing 747-136/236Bs, 44 Boeing 737-236s, 14 Lockheed L.1011-1/200 TriStars, 26 BAC One-Eleven 400/500s, five BAe 748-2Bs.

**British Caledonian Airways (B.Cal) (BR)** (1961)
Head Office: Caledonian House, County Oak, Crawley, West Sussex
Fleet: Two Boeing 747-200Bs, three A310 Airbus, nine McDonnell Douglas DC-10-30s, 13 BAC One-Eleven 500s. On order: seven A320 Airbus.

**British Caledonian Charter Ltd (JB)** (1984)
Head Office: Gatwick Airport, Crawley, West Sussex
Fleet: Two McDonnell Douglas DC-10-10s.

**British Emerald Airways** (1983)
Head Office: 56 Upper Queen Street, Belfast 1, Northern Ireland
Fleet: One McDonnell Douglas DC-10-40 (proposed).

**British Island Airways (KD)** (1982)
Head Office: Apollo House, Church Road, Lowfield Heath, Crawley, West Sussex RH11 0BQ
Fleet: Eight BAC One-Eleven 400/500s, one Piper PA-23-250C Aztec.

**British Midland Airways Ltd (BD)** (1938)
Head Office: East Midlands Airport, Castle Donington, Derbyshire DE7 2SB
Fleet: Eight McDonnell Douglas DC-9-14/32s, five Fokker F.27-200 Friendships, two Shorts 360s, nine Vickers V.800 Viscounts.

**Brown Air Services Ltd (BZ)** (1983)
Head Office: Wharfedale House, Harrogate Road, Pool-in-Wharfedale, Otley, West Yorkshire LS21 2RZ
Fleet: One Cessna 441 Conquest.

**Brymon Airways Ltd (BC)** (1972)
Head Office: City Airport, Crownhill, Plymouth, Devon PL6 8BW
Fleet: Three de Havilland Canada DHC-7s, two DHC-6 Twin Otter 300s.

**Channel Express (Air Services) Ltd (LS)** (1978)
Head Office: States Airport, Guernsey, Channel Islands
Fleet: Four Handley Page HPR.7 Herald 209s, one Partenavia P.68B Victor.

**Connectair Ltd (AX)** (1984)
Head Office: Vallance Byways, Lowfield Heath Road, Charlwood, Horley, Surrey RH6 0BT
Fleet: One Embraer EMB-110P1 Bandeirante.

**Dan-Air Services Ltd (DA)** (1953)
Head Office: Bilbao House, 36-38 Broad Street, London EC2 1NH
Fleet: Four BAe 146-100s, 1 Boeing 737-300, six Boeing 737-200s, 10 Boeing 727-200s, 18 BAC One-Elevens, 15 BAe 748s.

**Elan Air** (1984)
Head Office: East Midlands Airport, Castle Donington, Derbyshire DE7 2SB
Fleet: Two Armstrong Whitworth AW.650 Argosy 101s, one Handley Page HPR.7 Herald 401s.

**Euroair Transport Ltd** (1977)
Head Office: Room 109, Atlantic House, Gatwick Airport, Horley, Surrey RH6 0NP
Fleet: One BAe 125-1, eight Embraer EMB-110P1 Bandeirantes, one Piper PA-23-250 Aztec, and two Embraer Brasilias on order.

**Guernsey Airlines Ltd (GE)** (1977)
Head Office: States Airport, La Villiaze, Guernsey, Channel Islands
Fleet: One Vickers V.806 Viscount, two Shorts 330s.

**HeavyLift Cargo Airlines Ltd (NP)** (1978)
Head Office: Stansted Airport, Stansted, Essex
Fleet: Five Shorts SC.5 Belfasts, one Canadair CL-44-0 Guppy.

**Highland Express Ltd** (1984)
Head Office: Prestwick Airport, Ayrshire, Scotland KA9 2PL
Fleet: Three Lockheed L.1011 TriStars (proposed).

**Janus Airways Ltd** (1982)
Head Office: Lydd Airport, Romney Marsh, Kent TN29 9UP
Fleet: One Handley Page HPR.7 Herald 214.

**Jersey European Airways (JY)** (1978)
Head Office: States Airport, Jersey, Channel Islands
Fleet: Two de Havilland Canada DHC-6 Twin Otter 300s, one Embraer EMB-110P2 Bandeirante, one Shorts 330.

**Loganair Ltd (LC)** (1962)
Head Office: St Andrews Drive, Glasgow Airport, Paisley, Scotland PA3 2TG
Fleet: Two Shorts 360s, five de Havilland Canada DHC-6 Twin Otter 300s, six Britten-Norman BN-2A Islanders.

**London European Airways (UQ)** (1983)
Head Office: Administration Building, Percival Way, Luton Airport, Luton, Bedfordshire LU2 9NT
Fleet: One Vickers V.800 Viscount, one BAe Jetstream 31.

***Top:***
**Birmingham Executive's SAAB-Fairchild SF-340A photographed just prior to delivery in January 1985. It received British registration G-BSFI and is the first of its type to be registered in Britain.**
*SAAB-Fairchild*

***Above:***
**Birmingham Executive Airways also uses three Jetstreams on its European services.**
*BAe via Mike Hooks*

*Top:*
**Shorts 360 G-BMAJ is one of two operated by British Midland on scheduled services between London and East Midlands and Birmingham.** *BMA*

*Above:*
**Brown Air utilises a Cessna 441 Conquest on scheduled weekday services from Leeds/Bradford to Oslo and Frankfurt.** *Brown Air Services*

**Manx Airlines Ltd (JE)** (1982)
Head Office: Isle of Man (Ronaldsway) Airport, Ballasalla, Isle of Man
Fleet: Two Shorts 360s, one de Havilland Canada DHC-6 Twin Otter 300 (leased), one Vickers V.800 Viscount (leased).

**Metropolitan Airways Ltd (RD)** (1982)
Head Office: Bournemouth Hurn Airport, Christchurch, Dorset BH23 6DF
Fleet: Four Shorts 330s.

**Monarch Airlines Ltd (OM)** (1967)
Head Office: Luton Airport, Luton, Bedfordshire LU2 9NT
Fleet: Four Boeing 757-200s, four Boeing 737-200s. On order: Five Boeing 737-200s.

**Orion Airways Ltd (KG)** (1980)
Head Office: East Midlands Airport, Castle Donington, Derbyshire DE7 2SA
Fleet: Three Boeing 737-300s, six Boeing 737-200s. On order: Two 737-300s.

**Spacegrand Ltd (XF)** (1979)
Head Office: Blackpool Airport, Blackpool, Lancashire FY4 2QS
Fleet: three de Havilland Canada DHC-6 Twin Otter 300s, one Piper Navajo Chieftain, one Shorts 330.

**Tradewinds Airways Ltd (IK)** (1968)
Head Office: Timberham House, Gatwick Airport, Horley, Surrey RH6 0EY
Fleet: Three Boeing 707-320Cs.

**Virgin Atlantic Airways (VS)** (1984)
Head Office: 7th Floor, Sussex, House, High Street, Crawley, West Sussex RH10 1BZ
Fleet: One Boeing 747-200B, one Vickers Viscount (leased).

**British Caledonian Commuter** (1982)
Caledonian House,
County Oak,
Crawley,
West Sussex.

**British Commuter Airlines Association** (1984)
C/o Metropolitan Airways,
Bournemouth Hurn Airport,
Christchurch,
Dorset BH23 6DF.

**Gatwick Regional Airlines Association** (1984)
Viscount House,
Southend Airport,
Essex.

***Below:***
**G-AXOX is one of a fleet of One-Elevens operated by British Island Airways on inclusive-tour flights and contract services.** *BIA*

***Top:***
**Connectair uses a single Embraer Bandeirante, G-BNOC, on a commuter service between Gatwick and Antwerp in Belgium.** *Connectair*

***Above:***
**An historic photo from the birth of British Caledonian Commuter Services. Genair was the first airline to join but ceased operations in 1984. One of its Shorts 330s, G-BKIE, is pictured here at Gatwick Airport in company with one of B.Cal's DC-10-30s, G-BHDH.** *British Caledonian*

***Top:***
**The Shorts 330 is a favourite aircraft on UK commuter routes, and those in service include Metropolitan Airways' G-METO. The airline operates four of the type.** *John Cook*

***Above:***
**Euroair's mixed fleet used to include four Viscounts, one of which was available for charter in a VIP configuration for 36 passengers.** *Euroair*

***Top and Centre:***
**Two participants in the B.Cal Commuter system, both based in the Channel Islands, are Jersey European Airways which utilises Shorts 330s and de Havilland Canada Twin Otters, and Guernsey Airlines with Vickers Viscounts.**
*JEA/Guernsey Airlines*

***Above:***
**Elan Herald G-BEYF is operated on overnight parcel services centred on East Midlands Airport.**
*John Cook*

*Below:*
**A Datapost Bandeirante at Gatwick Airport, appropriately registered G-POST.** *John Cook*

***Below:***
**The big and unusual — HeavyLift's Short SC.5 Belfast and Canadair CL-44-0 Guppy.** *HeavyLift*

Virgin

LOGANAIR

HeavyLift
CARGO AIRLINES

BIA
British Island Airwa

TRADEWINDS
AIRWAYS

air UK

Aerotime

British Caledonian

ELAN

# British Air Taxis – Fly anywhere, anytime

British governments have never been very favourably disposed towards civil aviation, and the air taxi 'minnows' have suffered more than most from their attitude and that of the public, which viewed the air taxi as an unsafe and uncomfortable form of transport with exorbitant prices to boot. That the air taxi has not only survived against such overwhelming odds but grown in stature in the last 20 years to play a vital role in the country's air transportation system is due in no small measure to the entrepreneurial resilience of the individual operators and the work of the Air Transport Operators Association (ATOA). Through their joint efforts, air taxis now operate to the same high standards as the major scheduled airlines and although there still remains some lingering prejudice and ignorance, their manifold advantages are being recognised and accepted on an increasing scale. Over half a million passengers are now carried annually, utilising nearly 400 aircraft, and some 213 holders of Air Operators Certificates flew 137,000 hours in 1984.

**Air Transport Operators' Association**

ATOA was formed in 1967 as the Air Taxi Operators' Association to promote and protect the interests of a growing number of general aviation companies, and give a helping hand to develop their services. Since then it has gone from strength to strength and in 1983 adopted the new title to reflect its expanding membership which today includes 66 leading air taxi, third-level airline and other aviation-related companies.

A non-profit making business association, ATOA provides a centre where matters of common concern are considered on behalf of its membership, giving attention to such perennial problems as escalating charges and fuel costs, international regulations, security charges, noise legislation and airport and airfield facilities. Where necessary, representations are made to the Government, regulatory bodies or other organisations.

The Association, in partnership with the Civil Aviation Authority, has set high standards and codes of practice, in which flight safety is a primary consideration and membership is only available after formal application. The applicant has to prove able to meet all ATOA's requirements and those specified by legislation.

The association is represented on various committees and working groups including:

British Civil Aviation Standing Conference (Executive Committee).
BAA Consultation Groups: charges and airport development.
CAA-NATMAC and Airworthiness Consultative Committees.
Customs Consultative Committees.
Department of Transport: Business Aviation Working Group and Aircraft Noise Policy Group.
General Aviation: Safety and Fuel Committee.

On behalf of its membership ATOA is currently looking into airport facilities and developments; flight time limitations; CAA, airports, Eurocontrol and customs charges; and fuel duty, conditions and costs in the UK and Europe. Air ambulance services are being reviewed with the Department of Health, and an ATOA Standard Operations Manual is under preparation. Liaison with government departments, BAA, CAA, fuel suppliers, airports and other associations is an on-going part of its activities.

**Made-to-Measure Service**

Fast, flexible and first-rate are only some of the attributes of a successful air taxi operation. For the businessman especially it eliminates time-consuming surface travel to and from major airports, check-in and check-out hassles and expensive overnight stays, and instead offers convenient and comfortable direct point-to-point service in modern aircraft tailor-made to suit his needs.

A comprehensive network of over 200 airfields in the United Kingdom, supplementing those airports which are linked by scheduled passenger services, enables passengers to reach any business or leisure area by the shortest routes. Add to these another 1,300 airfields conveniently placed throughout Europe, providing direct and easy access to most continental locations, and the advantages become immediately obvious.

Aircraft operated include the latest designs of twin-engined executive jets, turboprops and

*Below:*
**The sophistication and comfort of today's business and air taxi aircraft are vividly illustrated in these photos of Beech King Air F90 G-BHUS.** *Eagle*

FLAP

# DISTRIBUTION OF AIR TAXI OPERATORS

***Above:***
**The McAlpine Aviation facilities at Luton International Airport, and one of the company's BAe 125-700s, G-BEFZ. McAlpine is one of the world's foremost executive charter companies and fixed based operators (FBO). It has won the best overseas FBO award from *Professional Pilot* in 1982, 1983 and 1984.** *McAlpine*

***Above:***
**Two American business jets are Townsend Thoreson's Cessna 550 Citation II G-FERY, and the first and to date only Gulfstream II on the British register, G-OVIP. The latter is operated by Marine & Aviation Management International.**
*John Cook/MAM*

***Top:***
**Trislander G-BDOS is operated by Kondair for XP Express Parcel Systems.** *John Cook*

***Above:***
**Woodgate Aviation is Northern Ireland's premier air taxi company based at Belfast International with a fleet of Piper Aztecs, including G-AZRG. A second base is maintained on the Isle of Man.**
*Woodgate Air Services*

# GROSVENOR AVIATION SERVICES LIMITEI

McAlpine Helicopters Limited

baa Blackpool Air Charter

MIDAS

merlix air LIMITED

Marine & Aviation Management International

***Above:***
**CEGA Aviation operates a twin-engined fleet for air taxi and charter work from Goodwood Airfield in Sussex. A Beechcraft 65-E90 King Air is pictured.**
*CEGA*

*Top:*
**Flagship of the NEA fleet is Learjet 35A G-LEAR. It can carry eight passengers in executive jet comfort at heights up to 13,500 m and speeds of 800km/hr; its non-stop range is 3,250km. The Learjet can also be fitted out for ambulance duties where there is ample accommodation for medical staff and equipment, and up to two stretcher cases. Northern Executive Aviation, based at Manchester, is one of the oldest air taxi operators and a founder member of ATOA.** *NEA*

*Above:*
**Ace Aviation's Piper. Navajo C/R G-JAKY is based at Glasgow Airport. The company also operates Navajo Chieftain G-NIKY, on air taxi and charter work.** *BAe*

***Above:***
**Northair, based at Leeds/Bradford, provides an executive charter service together with film work, aerial photography and air ambulance services. The company is the UK's only full-line Cessna dealer, and its fleet consists of various types, including the Cessna 425 Conquest I (*top*), 421 Golden Eagle, and the 441 Conquest II (*above*).** *Northair*

***Top:***
**The eight-seat Cessna 414A Chancellor G-MLCS is operated by Mountleigh Air Services on charter flights from Leeds/Bradford Airport.** *Mountleigh*

***Above:***
**Cessna 404 Titan G-EXEX is operated by Hubbardair from Norwich Airport.** *Hubbardair*

*Below:*
**Skycabs operates a Piper Seneca II, G-PLUS, on a charter service from the Channel Islands.** *Skycabs*

piston-engined types, many of which are pressurised to allow flights at higher altitudes for passenger comfort and economy. Available seating capacity spans a whole range from three to 36 passengers configured in spacious executive layouts or high-density 'mini-airline' arrangements. Some air taxis also operate helicopters which have the added advantage of almost unrestricted access to most points in the United Kingdom. The newest high speed executive jets with ranges exceeding 3,000km enable operations further afield and especially to North Africa and the Near and Middle East.

Costs are based upon flying hours and include the whole passenger or freight capacity of the aircraft which can be matched to clients' needs from the range available. Many factors play a part in arriving at the final cost of chartering such as the price of fuel, airport charges, overnight waiting time and type of aircraft, but the economics can readily be illustrated when comparing with scheduled air or other forms of transport.

Passenger transportation is by no means the only service the 75 or so companies provide in the United Kingdom. Carriage of urgent and sensitive freight, aerial survey and photography, and aircraft maintenance and sales are only some of the other facilities available. Quick response, flexibility and privacy also make air taxis ideal for air ambulance operations, and some also offer scheduled services providing convenient regular airline transport over established routes.

---

*ADDRESSES OF AIR TAXI OPERATORS IN THE UNITED KINGDOM*

**Ace Aviation Ltd** (1983)
Head Office: St Andrews Drive, Glasgow Airport, Paisley, Strathclyde, Scotland PA3 2SU
Fleet: One Piper PA-31-325 Navajo C/R, one Piper PA-31-350 Navajo Chieftain

**Aerotime Ltd** (1979)
Head Office: Building 98, General Aviation Terminal, Edinburgh Airport, Edinburgh, Scotland EH12 9DN
Fleet: One Beechcraft 65-C90 King Air

**Air Atlantique/General Aviation Services Ltd (DG)** (1969)
Head Office: States Airport, St Peter, Jersey, Channel Islands
Fleet: Six Douglas DC-3s, one Beechcraft 65-C90 King Air, one Cessna 421C Golden Eagle, three Cessna 310s

**Air Bristol** (1982)
Head Office: Hangar 31, Bristol Airport, Bristol, Avon BS19 3DY
Fleet: One Cessna 310R, one Piper PA-31-350 Navajo Chieftain

**Air Charter Scotland Ltd** (1965)
Head Office: Executive Lounge, Hangar 12, Glasgow Airport, Paisley, Scotland PA3 2TF
Fleet: Six Piper PA-31-350 Navajo Chieftain, four Piper PA-23-250 Aztecs

**AirFer Ltd** (1981)
Head Office: Regent Street, Newtownards, County Down, Northern Ireland
Fleet: One Piper PA-31-350 Navajo Chieftain, one Piper PA-23-250 Aztec

**Air Foyle Ltd (UP)** (1978)
Head Office: Luton Airport, Halcyon House, Luton, Bedfordshire LU2 9LW
Fleet: One Piper PA-31-310 Turbo Navajo, three Piper PA-23-250 Aztecs on order, one Piper PA-18A-150 Super Cub. Three Edgley Optica EA7s on order.

**Air Furness Ltd**
Head Office: Walney Island Airport, Barrow-in-Furness, Cumbria LA14 3YJ
Fleet: One Piper PA-23-250E Aztec

**Air Kilroe Ltd**
Head Office: Manchester International Airport, Hangar 520, Southside, Altrincham Road, Styal, Wilmslow, Cheshire
Fleet: One Beechcraft 200 Super King Air, one Piper PA-31-310 Turbo Navajo, two Piper PA-23-250 Aztecs

**Air London (Executive Travel) Ltd (GG)** (1960)
Head Office: Gatwick Airport South, Horley, Surrey RH6 0PG
Fleet: Two Piper PA-23-250 Aztecs

**Air Navigation & Trading Ltd** (1947)
Head Office: Blackpool Airport, Blackpool, Lancashire FY4 2QS
Fleet: Two Piper PA-23-250 Aztecs

**Air Orkney Ltd** (1981)
Head Office: 21 Bridge Street, Kirkwall, Orkney, Scotland KW15 1HQ
Fleet: One Britten-Norman BN-2A Islander

**Air Peterborough Ltd** (1981)
Head Office: Peterborough Business Airfield, Holme, Cambridgeshire PE7 3PK
Fleet: Two Piper PA-30-160B Twin Comanches

**Air Swansea Ltd**
Head Office: Terminal Building, Swansea Airport, Swansea SA2 7JU
Fleet: One Piper PA-31-310 Turbo Navajo

**Air Service Training Ltd** (1982)
Head Office: Perth Aerodrome, Perth, Scotland PH2 6NP
Fleet: Six Cessna 310s

**Air Transport Operators' Association** (1967)
Clembro House
Weydown Road
Haslemere
Surrey GU27 2HR

**Air Wight Ltd**
Head Office: Terminal Building, Bournemouth-Hurn Airport, Christchurch, Dorset BH23 60B
Fleet: One Britten-Norman BN-2A Islander, one Piper PA-23-250E Aztec

**Albion Aviation** (1980)
Head Office: Worth Corner, Pound Hill, Crawley, Sussex RH10 4SL
Fleet: One Cessna 550 Citation II

**Aly Aviation Ltd** (1967)
Head Office: Henstridge Airfield, Henstridge, Somerset BA8 0TF
Fleet: One Partenavia P.68C Observer, one Hughes 369C, one Hughes 269C

**ATS Air Charter Ltd** (1974)
Head Office: Blackbushe Airport, Camberley, Surrey GU17 9LG
Fleet: One BAe 125-403B, one Piper PA-31-350 Navajo Chieftain, two Piper PA-23-250 Aztecs

**Aviation Beauport Ltd** (1969)
Head Office: States Airport, St Peter, Jersey, Channel Islands
Fleet: One Cessna 421C Golden Eagle, one Piper PA-31-350 Navajo Chieftain

**Aviation West** (1982)
Head Office: Terminal Building, Bristol Airport, Bristol, Avon BS19 3DY
Fleet: Two Britten-Norman BN-2A Islanders, one Piper PA-23-235 Apache, one Cessna 414A Chancellor, two Aerospatiale AS.350B Ecureuils

**Avon Air Services**
Head Office: Bristol Airport, Bristol BS19 3DP
Fleet: One Beechcraft 95-B55 Baron, two Cessna 182P Skylanes

**BCA (Aviation) Ltd**
Head Office: Expedier House, Portsmouth Road, Hindhead, Surrey GU26 6TJ
Fleet: One Cessna 421C Golden Eagle, one Agusta A109A

**Beecham International Aviation Ltd** (1972)
Head Office: No 2 Maintenance Area, London Heathrow Airport, Southside, Hounslow, Middlesex
Fleet: One BAe 125-600B

**Birmingham Aviation Ltd** (1974)
Head Office: Hangar Road, Birmingham Airport, Birmingham B26 3QN
Fleet: One Piper PA-23-250 Aztec, one Piper PA-28R-200 Cherokee Arrow II, two Reims/Cessna 172 Skyhawks, four Reims/Cessna F152s, one Reims/Cessna FRA150

**Blackpool Air Charter Ltd**
Head Office: Blackpool Airport, Blackpool, Lancashire FY4 2QS
Fleet: One Piper PA-23-250 Aztec

**Bryan Aviation (Leasing & Management) Ltd** (1981)
Head Office: Building 226, Norwood Crescent, Heathrow Airport, Hounslow, Middlesex
Fleet: One BAC 111-301AG

**Business Air Travel Ltd** (1957)
Head Office: Lydd Airport, Romney Marsh, Kent TN29 9QN
Fleet: One Piper PA-23-250 Aztec, three Piper PA-34-200 Senecas

**Cabair Air Taxis Ltd** (1968)
Head Office: Elstree Aerodrome, Borehamwood, Hertfordshire WD6 3AW
Fleet: One Piper PA-31-350 Navajo Chieftain, one Piper PA-31T Cheyenne, one Piper PA-31-310 Turbo Navajo, six Gulfstream American GA7 Cougars. On order: Three Edgley EA7 Opticas

**Casair Aviation Ltd (KC)** (1972)
Head Office: Terminal Buildings, Tees-side Airport, Darlington, Durham DL3 1LU
Fleet: Two Cessna 404 Titans, one Piper PA-34-200T Seneca II, one Piper PA-23-250 Aztec

**CEGA Aviation Ltd (CN)** (1973)
Head Office: Goodwood Airfield, Chichester, West Sussex PO18 0PH
Fleet: One Beechcraft 65-E90 King Air, one Piper PA-34-200 Seneca

**Central Air Services (Air Envoy Ltd)** (1968)
Head Office: Birmingham Airport, Birmingham B26 3QB
Fleet: One Cessna 500 Citation I, one Cessna 421C Golden Eagle, one Piper PA-23-250 Aztec

**Colt Executive Aviation Ltd** (1978)
Head Office: Staverton Airport, Cheltenham, Gloucestershire GL5 6SR
Fleet: One Mitsubishi MU-300 Diamond 1, one Beechcraft 65-C90 King Air

**Continental Flight Services Ltd** (1976)
Head Office: Southampton Airport, Eastleigh, Hampshire SO9 1FQ
Fleet: One Piper PA-31-310 Turbo Navajo, one Piper PA-23-250 Aztec

**County Air Services Ltd**
Head Office: Headcorn Aerodrome, Shenley Farms, Headcorn, Kent TN27 9HX
Fleet: One Piper PA-31-310 Turbo Navajo

**Crest Aviation (JF)** (1983)
Head Office: Biggin Hill Airport, Biggin Hill, Westerham, Kent
Fleet: One Embraer EMB-110P1 Bandeirante. One Piper PA-31-350 Navajo Chieftain

**Directflight Ltd**
Head Office: Hunworth, Melton Constable, Norfolk NR24 2AA
Fleet: One Partenavia P.68B Victor

**Eagle Air Ltd** (1978)
Head Office: Cotworth Park, Ascot, Berkshire
Fleet: One Beechcraft 200 Super King Air, one Beechcraft 65-C90 King Air

**Eastern Air Executive Ltd** (1969)
Head Office: Sturgate Aerodrome, Upton nr Gainsborough, Lincolnshire DN21 5PA
Fleet: One Piper PA-31-350 Navajo Chieftain, one Cessna 421C Golden Eagle, two Piper PA-23-250 Aztecs, one Bell 47

**Elmdon Aviation Ltd** (1974)
Head Office: Birmingham Airport, Birmingham B26 3QW
Fleet: One Cessna 500 Citation, one Cessna 414A Chancellor

**Executive Airtours**
Head Office: Manchester International Airport, Hangar 522, Southside, Altrincham Road, Styal, Cheshire
Fleet: Two Piper PA-23-250 Aztecs, one Beechcraft 76 Duchess

**Expressflight Ltd** (1980)
Head Office: Southend Airport, Southend-on-Sea, Essex SS2 6UN
Fleet: One Piper PA-31-310 Turbo Navajo, one Cessna 404 Titan, one Cessna 421B Golden Eagle, one Piper PA-23-250E Aztec, one Piper PA-39 Twin Comanche C/R, one Piper PA-30-160 Twin Comanche, one Cessna 421B Golden Eagle, one Piper PA-23-250 Aztec, one Piper PA-30-160B Twin Comanche and one Piper PA-31-310 Turbo Navajo

**Fairflight Ltd** (1968)
Head Office: Biggin Hill Airport, Westerham, Kent TN16 3BN
Fleet: One Cessna 500 Citation I, two Shorts 360s, four Shorts 330s

**Falcon Jet Centre Ltd (FN)** (1974)
Head Office: No 2 Maintenance Area, London Heathrow Airport, Hounslow, Middlesex TW6 3AF
Fleet: One BAe 125-3B, one Dassault Falcon 20F

**Gatwick Air Taxis Ltd** (1980)
Head Office: Broadwater House, Chailey, Sussex BN8 4JE
Fleet: One Cessna 425 Corsair

**Grosvenor Aviation Services Ltd** (1981)
Head Office: Manchester International Airport, Wilmslow, Cheshire SK9 4LL
Fleet: Two Partenavia P.68B Victors, one Piper PA-31-350 Navajo Chieftain

**Hamilton Aviation Ltd** (1978)
Head Office: 4 Wests Centre, Bath Street, St Helier, Jersey, Channel Islands
Fleet: One Partenavia P.68C Victor

**Harvest Air Ltd**
Head Office: Municipal Airport, Southend-on-Sea, Essex
Fleet: Two Douglas DC-3s, 10 Britten-Norman BN-2A Islanders, one Piper PA-23-250 Aztec

**Hubbardair Ltd** (1980)
Head Office: Norwich Airport, Norwich, Norfolk NR6 6EL
Fleet: One de Havilland Canada DHC-6 Twin Otter 300, three Cessna 404 Titans, two Partenavia P.68B Victors, one Piper PA-31-310 Turbo Navajo

**IDS Aircraft Ltd (IS)** (1968)
Head Office: 93 Straight Road, Old Windsor, Buckinghamshire SL4 2SE
Fleet: Four Cessna 550 Citation IIs, one Cessna 500 Citation

**Interflight** (1974)
Head Office: Blackbushe Airport, Camberley, Surrey GU17 9LQ
Fleet: Three Piper PA-31-350 Navajo Chieftains, one Piper PA-23-250 Aztec, one Cessna 404 Titan

**Isles of Scilly Skybus Ltd**
Head Office: Land's End Aerodrome, St Just, Cornwall TR19 7RL
Fleet: One Britten-Norman BN-2A Islander

**Keenair Services Ltd** (1963)
Head Office: Hangar No 4, Liverpool Airport, Speke, Liverpool L24 8QQ
Fleet: One Piper PA-31-350 Navajo Chieftain, one Piper PA-23-250 Aztec, one Reims/Cessna F172N Skyhawk

**Kondair** (1977)
Head Office: Building 168, Stansted Airport, Essex CM24 8QW
Fleet: One Fokker F.27-600 Friendship, three Britten-Norman BN-2A Mk III Trislander, one Cessna 404 Titan, one Piper PA-31-310 Turbo Navajo, one Piper PA-23-250 Aztec

**Marine & Aviation Management International** (1973)
Head Office: PO Box 125, Heathrow Airport, Hounslow, Middlesex TW6 3BT
Fleet: One Gulfstream II, five BAe 125s (various models)

**McAlpine Aviation Ltd (RM)** (1947)
Head Office: Luton International Airport, Luton, Bedfordshire LU2 9NT
Fleet: Two BAC One-Eleven 400s, seven BAe 125s (various models), one Gates Learjet 35A, one Cessna 550 Citation II, four BAe 3102 Jetstream 31s, one Cessna 441 Conquests, two Cessna 421B/C Golden Eagles

**Merlix Air Ltd**
Head Office: Fairoaks Airport, Chobham, Woking, Surrey GU24 8HU
Fleet: One Piper PA-31-350 Navajo Chieftain, one Piper PA-23-250 Aztec

**Midas Airways** (1977)
Head Office: Cambridge Airport, Newmarket Road, Cambridge CB5 8RX
Fleet: Two Piper PA-23-250 Aztecs

**Midland Air Taxis Ltd**
Head Office: Halfpenny Green Airport, Stourbridge, West Midlands
Fleet: One Piper PA-34-200 Seneca II

**Mountleigh Air Services Ltd** (1981)
Head Office: Leeds/Bradford Airport, Yeadon, Leeds, Yorkshire LS19 7T2
Fleet: One Cessna 414A Chancellor

**National Airways (MO)** (1980)
Head Office: Elstree Aerodrome, Borehamwood, Hertforshire WD6 3AW
Fleet: One Beechcraft 200 Super King Air, three Beechcraft 65-C90 King Airs, one Piper PA-34-200 Seneca

**Northair Aviation Ltd (NT)** (1964)
Head Office: Leeds/Bradford Airport, Yeadon, Leeds, Yorkshire LS19 7YG
Fleet: One Cessna T303 Crusader, three Cessna 425 Corsairs, two Cessna 441 Conquest IIs, three Cessna 421C Golden Eagles, one Cessna 414A Chancellor, one Cessna 340A

**Northern Executive Aviation Ltd (NV)** (1962)
Head Office: Hangar 522, Manchester International Airport, Wilmslow, Cheshire SK9 4LL
Fleet: One Gates Learjet 35A, one Piper PA-31-350 Navajo Chieftain, one Piper PA-31-310 Turbo Navajo, two Piper PA-23-250E Aztecs, one Piper PA-30-160 Twin Comanche. On order: Three Edgley EA7 Opticas

**Omega Air Ltd** (1981)
Head Office: Building 509, Biggin Hill Airport, Westerham, Kent TN16 3BN
Fleet: One Piper PA-31-310 Turbo Navajo

**Peregrine Air Services Ltd (PJ)** (1966)
Head Office: Aberdeen Airport East, Dyce, Aberdeen, Scotland
Fleet: Two BAe 3102 Jetstream 31s, two Beechcraft 200 Super King Airs, three Cessna 404 Titans, four Piper PA-23-250 Aztecs

**Skycabs** (1970)
Head Office: 8 Mulcaster Street, St Helier, Jersey, Channel Islands
Fleet: One Piper PA-34-200T Seneca II

**Skyguard Ltd**
Head Office: 72 High Street, Rickmansworth, Hertfordshire WD3 1AJ
Fleet: One Handley Page HPR.7 Herald 213, one Piper PA-31-350 Navajo Chieftain, one Beechcraft 95-58 Baron

**Sparrows Aviation** (1980)
Head Office: Bristol Airport, Bristol, Avon BS19 3DY
Fleet: One Piper PA-31-350 Navajo Chieftain

**Tayside Aviation Ltd** (1968)
Head Office: Dundee Airport, Riverside Drive, Dundee, Scotland DD2 1UK
Fleet: One Piper PA-23-250 Aztec

**TAL-Air (Thurston Aviation Ltd) (HZ)** (1955)
Head Office: London Stansted Airport, Stansted, Essex CM24 8QW
Fleet: One BAe 125-3B, two Beechcraft 200 Super King Air, one Beechcraft 65-E90 King Air, two Piper PA-31-310 Turbo Navajos, two Piper PA-23-250 Aztec, one Hughes 369D

**Topflight Aviation Co Ltd** (1978)
Head Office: Building D, Fairoaks Airport, Chobham, Woking, Surrey GU24 8HU
Fleet: One BAe-125-400B, three de Havilland DH.114 Heron 2Ds, two Piper PA-31-350 Navajo Chieftains, one Piper PA-23-250C Aztec

**Tragoair (Bodmin Aviation Ltd)** (1981)
Head Office: Bodmin Aerodrome, Cardinham, Bodmin, Cornwall PL30 4PU
Fleet: One Piper PA-23-250D Aztec

**Vernair Flying Services (VC)** (1967)
Head Office: Liverpool Airport, Speke, Liverpool L24 9HE
Fleet: One Beechcraft 200 Super King Air, two Beechcraft A100 King Air, two Beechcraft 65-C90 King Airs

**Welsh Airways Ltd**
Head Office: Haverfordwest Aerodrome, Haverfordwest, Dyfed, Wales SA67 4BN
Fleet: One Partenavia P.68B Victor, one Piper PA-23-250E Aztec

**Wessex Air Charter Ltd**
Head Office: Exeter Airport, Clyst Honiton, Exeter, Devonshire EX5 2BD
Fleet: One Piper PA-31-310 Turbo Navajo, one Piper PA-23-250E Aztec

**Westair Flying Services Ltd**
Head Office: Blackpool Airport, Blackpool, Lancashire FY4 2QX
Fleet: One Cessna 401

**Westward Airways (Lands End) Ltd** (1970)
Head Office: Land's End Aerodrome, St Just, Penzance, Cornwall
Fleet: One Cessna 337C Super Skymaster, one Reims/Cessna F172F Skyhawk, one Reims/Cessna FA152 Aerobat

**Woodgate Air Services Ltd** (1969)
Head Office: Belfast International Airport, Belfast, Northern Ireland
Fleet: Four Piper PA-23-250C/D Aztecs, one Piper PA-31-350 Navajo Chieftain

Mountleigh Air Services Limited

TAL-AIR

COLT

COLT EXECUTIVE AVIATION

BRYAN
AVIATION

McALPINE
McA
AVIATION

Crest Aviation Ltd.

TOP FLIGHT

ROUTAIR

NORTHAIR

AIR SERVICES IN WALES
WELSH AIRWAYS
LIMITED

FLY Eagle

Fairflight

Northern Executive Aviation Limited

KEA

DIRECTFLIGHT

CASAIR

Cega Aviation Limited

Aviation Beauport Limited

AB

# Britain's Helicopter Operations

**Britain's Helicopter Operations—** The helicopter is indisputably the most versatile aerial vehicle man's ingenuity has yet managed to devise. As with fixed wing aircraft, wars like those in Korea and Vietnam (in particular), have hastened its development, but it can be said with some justification that the greatest leap forward in recent times has been due, more than any other single factor, to the discovery of oil and gas in offshore locations. The hostile environment of the North Sea posed a challenge that called for new and exacting requirements, a challenge which was met by the manufacturers with considerable success.

When British European Airways formed its Helicopter Experimental Unit in 1947, it could not then have foreseen the vital and versatile role the rotor-wing craft would play in Britain 40 years later. And yet it did have men of vision in N. E. Rowe and Reggie Brie who perceived, if only vaguely, the enormous possibilities of the type and kindled the pioneering spirit in men such as Jock Cameron, Alan Bristow and Eric Brown, who became Britain's first rotary wing research test pilot.

After the war, production of helicopters in the UK was in the hands of three main companies, the Bristol Aeroplane Co, Fairey Aviation and Westland Aircraft. Whilst Westland concentrated on licence production of American machines, the other two experimented with several indigenous designs. Noteworthy in the years that followed were the two-seat Bristol 171 Sycamore, the first British helicopter to receive a certificate of airworthiness; the Type 173 Belvedere, the first twin-engined tandem-rotor helicopter; and the advanced Fairey Rotodyne which first flew in 1957 but foundered, as so many advanced aviation products, due to lack of government support.

## WESTLAND HELICOPTERS

Westland is now the only organisation engaged in helicopter design and manufacture in Great Britain though its origins go back to 1915 when Petters, an industrial company located at Yeovil, set up Westland Works to manufacture seaplanes for the Royal Navy.

The company entered the helicopter industry in 1947 by acquiring a licence to build the Sikorsky S-51 for which a production line was established in 1950. A total of 133 were produced, known as the Westland Dragonfly. The association with Sikorsky continued, resulting in the development and production of several more types, principally for military use, including the Sea King, Whirlwind and Wessex helicopters. By the time a new company, **Westland Helicopters Ltd**, was set up on 1 October 1966, Saunders-Roe and the helicopter divisions of Bristol Aircraft and Fairey Aviation had already been acquired making Westland the UK's largest rotary wing manufacturer.

### Westland 30

The company's first venture into the purely civil helicopter market came with the wide-bodied Westland 30, based on the successful Lynx helicopter of which more than 200 are in service. The Westland 30 (G-BGHF) made its maiden flight on 10 April 1979 and received its CAA Type Certificate for IFR operations on 30 March 1982, having logged over 800 hours of development flying. The first production aircraft was delivered to British Airways Helicopters in January 1982. A total of nine have been delivered to date to the following customers: British Airways Helicopters (three), Helicopter Hire (one), and abroad to Airspur Helicopters (four) and Omniflight/Pan Am (one) but a lack of further orders has put Westland into severe financial difficulties.

With the Westland 30, the company has succeeded in lifting the rotary wing vehicle out of the utility class, undoubtedly setting new standards for the future. Designed from the outset with the commuter and airline interline market in mind, Westland has come up with a helicopter which provides airliner style comfort not previously available in medium-sized, twin-engined helicopters. The cabin can be fitted for up to 19 passengers, and the large windows, ample headroom and central isle enhance the feeling of spaciousness. Aircraft noise and vibration have also been greatly reduced, and seats can be removed quickly for cargo operations. Role flexibility has been built in from the start and the Westland 30 is particularly suitable for offshore operations where its proven

***Top:***
**G-BGHF was the first Westland 30. It made its maiden flight in April 1979.** *Westland*

***Above:***
**A model of the 30-passenger EH101 three-engined helicopter being developed jointly by Westland and Agusta of Italy.** *Westland*

| | Westland 30 | EH 101 |
|---|---|---|
| *Dimensions* | | |
| Length | 14.23m | 15.85m |
| Cabin width | 1.98m | 2.39m |
| Rotor diameter | 13.31m | 18.59m |
| *Weights* | | |
| Maximum take-off | 5,806kg | 14,290kg |
| Useful payload | 2,088kg | 6,699kg |
| *Powerplant* | | |
| | 2×880kW Rolls-Royce Gem 60-3 turboshafts | 3×1,500kW General Electric CT7-8 turboshafts |
| *Performance* | | |
| Maximum cruising speed | 222km/hr | 296km/hr |
| Maximum range | 723km | 1,020km |

Rolls-Royce Gem engines offer twin-engined safety for over-water operation and all-weather capability. Safety and reliability in all conditions has of course been a prime consideration from the outset and the Westland 30 incorporates all the latest 'state of the art' technology.

In its role as an air ambulance it can take six stretcher cases and an equal number of seated casualties, and a VIP/Executive layout is offered with comfortable accommodation for 6-10 passengers. Provisions are also available for the installation of an external freight sling, rescue hoist, and specialised fire-fighting, search and rescue, medical and police role equipment.

## EH 101

In conjunction with Agusta, the prolific Italian manufacturer, Westland is developing a new 30-seat, long-range multi-purpose helicopter, the EH 101, aimed at a variety of both military and commercial applications. The programme is being managed by **E. H. Industries Ltd**, a company jointly controlled by Westland Helicopters and Agusta SpA, formed in June 1980 and funded by the British and Italian governments. The first flight is scheduled to take place in 1986, with deliveries to customers commencing before the end of the decade.

The three-engined EH 101 evolved from a British and Italian Naval requirement for a large-capacity helicopter with sufficient endurance to undertake its missions in adverse marine operating environments. Market research indicated that although offshore operations, logging, mines and onshore exploration and commuter services all present their own specific requirements, they were essentially very similar and could be combined in a single helicopter without significant compromise.

In civil use, up to 30 passengers can be carried in airline comfort over a distance of 550 nautical miles.

It is generally believed that some 15,000 helicopters will be required for civil use between now and the year 2000. With the W30 and EH 101, Westland is ready to take its share of the years ahead.

## The Aerial Workhorse

The military applications of the helicopter, always predominant in the conceptual design stage, are much publicised, but the civil and commercial uses in the United Kingdom are considerable and far more varied and interesting than one would imagine. A primary task is of course the lifting of heavy equipment, construction materials and such like to difficult and otherwise inaccessible locations. With its ability to hover almost motionless, and the advent of inertia-mounted cameras, the helicopter is now ideally suited to aerial photography, photogrammetry, surveys and mineral exploration, and there are a number of UK companies specialising in this type of work. Various authorities and companies use helicopters for routeing and inspection of pipelines and powerlines, and the police find them a useful vehicle for traffic control and law inforcement.

There are currently three scheduled helicopter services in Britain, and helicopters are also frequently chartered by those who can afford it (helicopters are expensive to operate) to whisk them to race meetings, golf tournaments and other sporting events. Agricultural spraying and seeding is another task where the helicopter, with its ability to fly low and slow, scores over fixed-wing aircraft, especially in the undulating terrain of vineyards and orchards. The merit of helicopters in an aerial ambulance role has been successfully demonstrated in the Alpine countries and has now become an established part of rotary wing operations in the UK, as have search and rescue (SAR) services which are undertaken partly by the military and partly by private contracted operators.

The protection of shipping in an island nation is also of vital importance, and the Trinity House Lighthouse Service and the Northern Lighthouse Board in Scotland employ helicopters for the monthly relief of lighthouse keepers and crews of light-vessels. Increasing use is made by television and film companies for outside news broadcasts and location filming.

**Helicopters over the North Sea**

Exploration of the North Sea's energy sources began in earnest in 1965 with the first gas strike off the East Anglian coast. The search moved northwards, and in 1969 the first oil was brought up off the northeast coast of Scotland.

The unfriendly environment makes the extraction of oil and gas extremely expensive, and their successful exploitation would probably not have been possible without the helicopter to provide offshore support services. These consist of ferrying rig crews to and from the mainland, the provision of spares and supplies, and inter-rig shuttle flights with underslung loads. In addition, there is a 24-hour availability requirement for emergencies. At the most intense phase, during construction and hook-up, once oil has been found, one rig may accommodate anything up to 500 people. Such large numbers, together with the ever-shortening work cycle which is now seven days on and seven days off, places a heavy demand on the helicopter.

There are presently a staggering 1,500 daily helicopter movements, and the concentration of activities in the major bases in the north and southeast conflict with military training zones. The worst problem, in the East Shetland Basin, has been eased by the creation of a Helicopter Flight Information Service Area (HFISA) and the use of the Ministry of Defence radar at Saxa Vord, while in East Anglia a helicopter corridor is in operation from the mainland to the rigs. There are also Helicopter Protected Zones (HPZs) around each rig complex in the North Sea.

The North Sea is one of the most inhospitable marine areas in the world. Weather conditions for much of the time are quite appalling, with wind speeds of 60kt, sea fog, and snow and ice in winter common occurrences. The weather is therefore the most vital factor in off-shore operations and only IFR capability and twin-engined safety is acceptable. Helicopters are now equipped with the proven Decca navigation system, and also include VOR (VHF Omni-Range), ADF (Automatic Direction Finder), weather radar, radio altimeter and VHF/UHF radio. In addition, major rigs are fitted with Non-Directional Beacons (NDBs) for the terminal approach phase. Helicopters are cleared to operate in wind speeds of up to 70kt, and in light icing conditions at temperatures down to −5°C. The CAA has also laid down minimum standards for helidecks, and the positioning of Helicopter Landing Officers on off-shore platforms to assist with operational matters is yet another facet to the UK's strict safety precautions.

There are two major helicopter bases in Scotland, located at Aberdeen Dyce Airport and Sumburgh Airport in the Shetlands, both of which have experienced phenomenal growth since 1969, due almost entirely to oil-related traffic. Helicopter passengers at Aberdeen alone are now in the region of 600,000 annually which, in terms of movements, makes it the third busiest airport in the United Kingdom.

There are presently about 120 UK helicopters flying on North Sea support services varying in capacity from 4-44 passengers, with payloads from 500kg to 10,000kg, and operational ranges up to 1,000km. The most popular type since the beginning has been the 26-seat Sikorsky S-61N which still makes up 37% of the total fleet. Other prominent participants are the Boeing BV234 Chinook, the Aerospatiale AS.332L Super Puma and the Sikorsky S-76A. The helicopter's outstanding record and reliability in the North Sea has been achieved by meticulous safety standards, rigorous training and continuous development of techniques and machines. The most experienced operator is **British Airways Helicopters Ltd**, having been the first company to set up a support base, at Beccles near Great Yarmouth, when the search for oil and gas started off East Anglia in 1964. As the search moved northwards, new bases were established at Aberdeen and Sumburgh, providing the growth with the oil industry which now brings in well in excess of 90% of its total revenue. The rest comes from scheduled services, air-sea rescue and heavy lift operations.

British Airways Helicopters has a long record of successful rotary wing innovation starting in 1950 with the world's first scheduled passenger service. In 1964 it introduced the Sikorsky S-61N into off-shore operation, and followed this up in 1982 with the long-range 44-seat Boeing-Vertol BV234 to offer direct links to the more distant platforms from the Scottish mainland. It was the

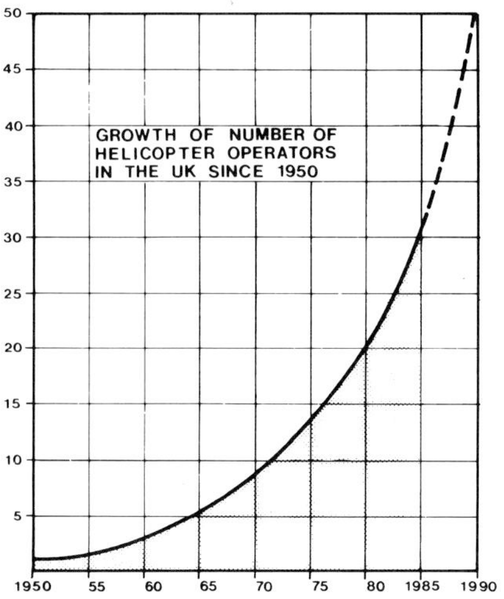


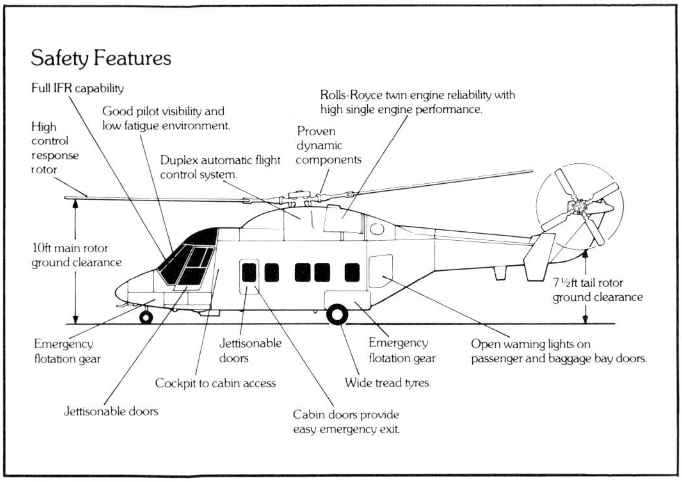

***Centre:***
**Offshore production workers disembark from British Airways Helicopters' Westland 30 G-BIWY after the flight from the shore to the platform.**
*Westland*

***Above:***
**British Airways Helicopters utilises four of the giant Boeing-Vertol Chinooks which can carry up to 44 passengers over long ranges.**
*British Airways Helicopters*

first company to secure CAA approval for IFR operations and to obtain clearance to fly in icing conditions, and in 1978 became the first civil operator in the world to install a highly sophisticated six-axis simulator, for its fleet of Sikorsky S-61Ns. Another for the Boeing Vertol 234 has since been added.

Each simulator incorporates the latest technology for its type and can reproduce virtually any weather, loading or operating conditions, including many situations, such as loss of tail rotor, which could not be carried out on the actual aircraft. The simulators are used for all mandatory pilot checks as well as conversion, emergency and route procedure training. British Airways Helicopters also provides simulator and classroom training facilities for civil operators from many countries and for military establishments, including the United Kingdom's Royal Air Force.

At the present time British Airways Helicopters operates a fleet of 17 S-61Ns, four BV234 Chinooks, three AS.332L Super Pumas, four S-76s and three Westland 30s. The numbers at each location change constantly but break down roughly as follows: Aberdeen — nine S-61Ns, three Super Pumas and four Chinooks; Sumburgh — three S-61Ns; Beccles — three S-76As and one Westland 30; Penzance — one S-61N; and Humberside — two Westland 30s. Additionally, two S-61Ns are based at Zhuhai in mainland China and at Porbander in India.

Utilisation of the Chinook averages out at 150 hours per month, and for the S-61N at 120 hours per month.

By far the most prolific operator in Britain's North Sea is **Bristow Helicopters Ltd**. The company, whose expertise in this type of work started in the Antarctic in 1951, has grown to become one of the world's largest and most experienced helicopter groups, operating in support of oil and gas exploration and production worldwide. Bristow's expertise also extends to all other types of helicopter activities including survey and exploration, executive charter, search and rescue, harbour pilots and ship's supplies transfers, pipeline inspection and public transport winching operations. Bristow Helicopters, through associated and subsidiary companies, is also active in many other parts of the world.

Bristow services over 100 rigs and platforms stretching from Great Yarmouth to the Magnus Field, some 250km northeast of Sumburgh in the Shetlands. Of the company's 200 helicopters, over 50 are operated in and around the North Sea from six bases, and include 24 Aerospatiale AS.332L Bristow Tigers, nine Bell 212s, seven Sikorsky S-76 Mk IIs and 14 Sikorsky S-61N Mk IIs. Bases are maintained at Aberdeen, Sumburgh, Unst, Newcastle, Great Yarmouth and offshore. Aberdeen provides the hub for most of Bristow's North Sea operations, where it has extensive administrative and maintenance facilities. An instrument training school is also operated at Great Yarmouth.

Bristow currently operates a dedicated SAR S-61N helicopter for HM Coastguard in Sumburgh, Shetland Islands, on behalf of the UK Department of Transport. This helicopter, like four Bell 212 helicopters operated on behalf of Shell Expro in the Brent Field, is equipped to provide a hover capability day or night irrespective of weather conditions. These five helicopters are the only civil helicopters in the world with such a capability.

The Bristow Group has flown in excess of two million hours, and carried more than 18 million passengers and one billion pounds of freight on over four million flights. The Group currently flies about 10,000 hours and carries 150,000 passengers per month.

In the UK, on offshore operations, Bristow Helicopters flew 65,000 hours and uplifted 883,000 passengers in 1984.

Well over a decade of service has been logged by **Bond Helicopters Ltd**, formerly North Scottish Helicopters Ltd and Management Aviation, which began North Sea operations in 1972. The company is particularly experienced in offshore work where it has logged over 200,000 helicopter hours, most of these in oil and gas exploration and production support. Its modern twin-engined offshore fleet ranges in capacity from four to 26 passengers and operates from three bases — Aberdeen, Peterhead and Sumburgh. Stationed at Aberdeen are the six Aerospatiale AS.332L Super Pumas and five Sikorsky S-76As, whilst the Peterhead complement includes Aerospatiale SA.365N Dauphin IIs and MBB Bölkow Bö.105Ds. No helicopters are presently based at Sumburgh.

The Super Puma and S-76A helicopters are used for oil crew changes making an average of 12 flights per day of approximately four-hour duration. The Dauphins and Bö.105Ds are employed on offshore shuttle and SAR duties, with the exception of those currently on a Ministry of Defence contract at Kyle of Lochalsh, Plockton, and on a contract for the Northern Lighthouse Board for crew changes and lighthouse supplies.

Although a relative newcomer to the North Sea, **British Caledonian Helicopters Ltd** has quickly established itself in this highly specialised market, and has won major contracts in the face of fierce competition. From its early days, BCHL's long term development plans were focused on the North Sea, and in 1980 the

***Top:***
**Winch training on Bristow's Sikorsky S-61N Mk II, G-BDOC, dedicated to search and rescue.** *Bristow*

***Above:***
**Aerospatiale AS.332L *Bristow Tiger*, G-TIGK, approaching Conoco's Murchison platform.** *Bristow*

***Top:***
**Bristow Bell 212 G-BJZS over the 'Forties Bravo' platform.** *Bristow*

***Above:***
**Bond Helicopters' two main types on the North Sea are the Aerospatiale SA.365N Dauphin II (*left*) and the AS.332L Super Puma (*right*).**
*Bond Helicopters/Allan Burney*

company transferred its operational base from Shoreham on the south coast to Aberdeen's Dyce Airport in Scotland where a modern operational and administrative centre was opened in late 1981 which today houses some 150 ground and flying staff working in four departments — operational, engineering, customer services and finance/administration. The BCHL maintenance facility is capable of accommodating up to 10 large helicopters at any one time.

Of its mixed fleet of Bell and Sikorsky helicopters, five are based at Aberdeen including three long range 18-seat Bell 214STs, one Sikorsky S-61N, and the fastest machine in the fleet, the Sikorsky S-76A. Ten to 12 flights are made each weekday, with rather less at the weekend. Daily utilisation varies due to the far flung locations of the oil companies' rigs, but averages out at about five hours per aircraft. The four S-61Ns stationed at BCHL's Sumburgh base operate some eight to 10 flights each day.

### Scheduled Services

The United Kingdom can currently boast three scheduled helicopter operations, the same as the rest of Europe together, and not far behind the United States. Historically though, Britain has no equal, even if early attempts were largely punctuated by brave tries and disappointing failures.

After setting up its Helicopter Experimental Unit in 1947, **BEA** began with night mail services under contract to the Post Office and went on to open the world's first scheduled helicopter service in June 1950 between Liverpool, Wrexham and Cardiff using four-seat Sikorsky S-51s. This was superseded in 1951 by a service between Birmingham and the London airports at Northolt and Heathrow, followed a year later by a London-Southampton link. Other routes were tried subsequently but all proved uneconomical and lasted only a few months.

These experiments highlighted two major deficiencies — the lack of twin-engined helicopters with suitably large capacity, and all-weather capability. The problems were largely overcome in March 1964 when BEA purchased the Sikorsky S-61N which immediately led to a highly successful and profitable operation between Land's End and the Isles of Scilly, taking over its own fixed-wing service flown with de Havilland Rapides. The service was inaugurated on 1 May 1964 from St Just (later transferred to a new heliport near Penzance railway terminus), and the 24-passenger S-61N carried nearly 48,000 passengers in its first year. The one millionth passenger was recorded in July 1979 and the service now generates some 90,000 passengers a year.

The Isles of Scilly lie about 45km southwest of Land's End. A flourishing tourist industry provides for most of the passengers during spring, summer and autumn and, due to its mild climate, the export of flowers accounts for a large proportion of winter traffic. During the busy summer days 12 return flights are scheduled daily with rather less in the winter months. Maintenance is carried out on Sundays on which day there is no service. A one-way journey takes 20 minutes and the Sikorskys (now fitted out for 34 passengers) operate at 98% regularity.

### The Gatwick Heathrow Airlink

A scheduled helicopter link service between the two London airports is provided by the British Airports Authority in conjunction with British Caledonian Airways. Ten round trips are scheduled each day, seven days per week, and the 15-minute flights are timed to coincide with the morning and late afternoon peak periods at both airports. The aircraft, a standard production Sikorsky S61N with seating for 26 passengers, is leased from British Caledonian Helicopters. Terminal facilities, including check-in desks and office accommodation, are provided by the BAA, and technical support and flight deck crews by British Caledonian Helicopters.

The idea of the Gatwick Heathrow Airlink was first mooted in 1976 when B.Cal and the British Airports Authority met to discuss the transfer of passengers between the two airports, a problem of common interest. A substantial number of the international passengers travelling into London are making interline connections, many of which necessitate transfer between Gatwick and Heathrow. Various proposals were investigated including more cars and buses, rail services, helicopters and STOL aircraft. Helicopters were chosen for their minimal impact on scarce runway slots and their short sector times compared with the lengthy fixed-wing routeing. Since the inauguration on 9 June 1978 by HRH the Prince of Wales, a total of 600,000 revenue passengers have been carried on this inter-airport link.

Whilst it is difficult to argue with the benefit to aviation of this service, a strong environmental lobby has nevertheless managed to cloud the future of Airlink by getting the Ministry of Transport to overturn the decision in 1983 to grant a new 10-year licence. This new ruling states that Airlink will have to cease four months after the completion of the M25 motorway, scheduled for August 1985. If Airlink is stopped the effect on both London airports will be of significant proportions. It is forecast that about one quarter of all Airlink passengers would not

***Top:***
**Bond Helicopters also operates MBB Bölkow Bö.105s for the Northern Lighthouse Board.**
*Bond Helicopters*

***Above:***
**British Caledonian Helicopters is the only company to use the long-range 18-seat Bell 214ST on North Sea operations. G-BKJD is pictured here at Aberdeen Dyce.** *British Caledonian Helicopters*

## UK SCHEDULED HELICOPTER ROUTES

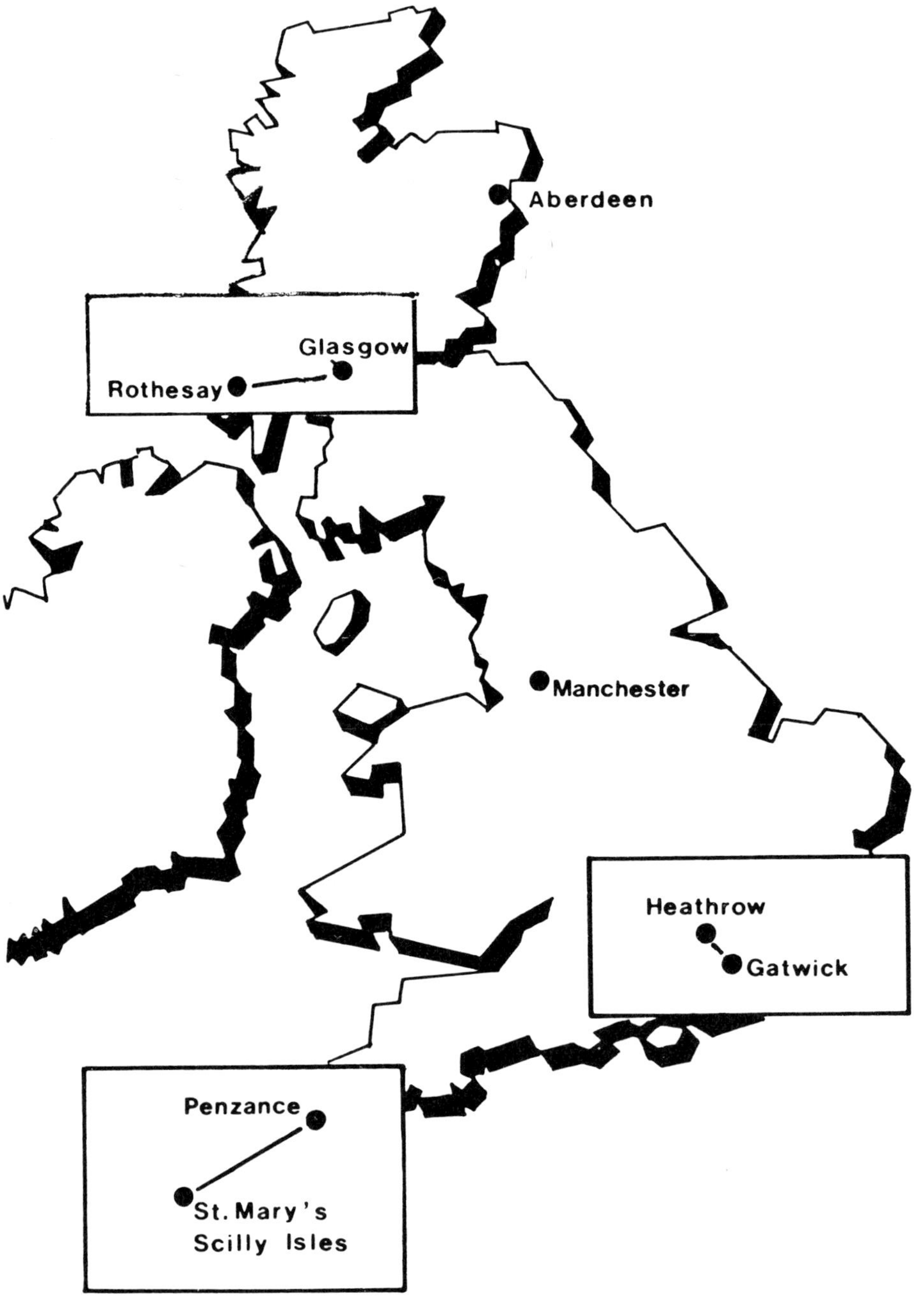

transit through London, but would use Paris, Amsterdam and Frankfurt instead.

Also in 1978, **Burnthills Aviation Ltd** put a five-passenger Bell JetRanger on to a five-times weekly schedule between Glasgow and Fort William, later increased to nine round trips per week. Fort William was selected in view of (a) its position in the West Highlands, (b) its unsuitability for fixed wing operations and (c) its development, both industrial and commercial, as a centre for employment on the Scottish West Coast. The service could not be considered self supporting and the first year's operation was supported by financial assistance from the Highlands & Islands Development Board, Highland Region and Lochaber District Council. Another route to Rothesay and Lochgilphead was inaugurated on 29 April 1981 and the inclusion of the West Coast resort of Oban as an optional stop on the Fort William route was commenced on 24 July 1981. The HIDB subsidy was unfortunately withdrawn, and the services to Lochgilphead, Oban and Fort William ceased on 31 March 1984. The Rothesay service remains, and is operated twice weekly with the financial risk being taken by Burnthills.

Since the commencement of services in December 1978, there have been many enquiries from regions, councils and communities (particularly in the West of Scotland) which have long and arduous travelling to the nearest centre of population and which would benefit from a direct air link, but no one has been able to provide the necessary finance. Among the most notable are Isle of Arran to Glasgow, Dunoon to Glasgow and Lochgilphead, Isle of Lismore to Glasgow and Oban, and Colonsay to Oban and Lochgilphead. The company is currently licensed to permit optional stops at Oban, Colonsay, Jura, Mallaig, Portree, Inverness, and Invergordon.

Even after 35 years, the use of helicopters on scheduled routes is still little more than in its infancy. Nevertheless, with the advent of bigger, quieter and much more comfortable machines, manufacturers are convinced that helicopters will play an increasing role particularly in airport to airport and airport to city centre connections over the next decade.

## British Helicopter Advisory Board (BHAB)

The interests of the UK helicopter fraternity are represented by the British Helicopter Advisory Board which was formed in January 1969 to act for an industry which, though growing at a fast pace, had until then no single authoritative voice to speak for it. The main objective of the BHAB is to promote the use of helicopters through liaison with relevant government departments, local authorities, public services and private industry. Exchange of information with similar foreign helicopter organisations is actively promoted. There are additionally five permanent working committees covering the main fields of the Board's activities — heliport, operations and air legislation, technical, charges and offshore operators. Other sub-committees are formed as and when necessary to deal with specialised problems.

The Board is a non-profit-making organisation and is financed by subscriptions from its members at rates which are scaled to the size of the operating capacity of the member. The membership today ranges from major manufacturers and commercial operators down to private owners and distributors, and encompasses some 87% of a total of 580 helicopters on the UK Civil Register.

The wider interests of helicopter operators in the European community are served by the **European Helicopter Association** which was established in 1981 on the instigation of the BHAB and is also headquartered in the UK. It works closely with the Air Transport Advisory Division of the European Economic Community and with the International Civil Aviation Organisation, contributing to ICAO's Helicopter Operations Panel through an Operations and Air Legislation Committee, which deals with all aspects of operational and regulatory problems.

The EHA also has a very active Offshore Committee which has been involved principally in North Sea operational matters, and is now expanding its interest to other European waters where oil and gas exploration is underway or likely in the future. It also has a Technical Committee with the goal of promoting Joint Airworthiness Requirements in the interests of standardisation. The EHA holds a two-day annual symposium in a European capital to discuss all aspects of helicopter operations and to foster closer liaison with civil airworthiness authorities.

---

*ADDRESSES*

**British Helicopter Advisory Board (BHAB)**
Knowles House, Cromwell Road, Redhill, Surrey RH1 1LW

**E.H. Industries Ltd**
Granville House, 132-135 Sloane Street, London SW1X 9BB

**European Helicopter Association (EHA)**
Knowles House (Room 204), Cromwell Road, Redhill, Surrey RH1 1LW

***Top:***
**Alan Mann Helicopters operates from Fairoaks Airport with a mixed fleet which includes three Agusta A109s. G-OAMH, though no longer in the fleet, illustrates this twin-engined Italian helicopter.** *John Cook*

***Above:***
**Air Hanson's Sikorsky S-76 Executive, registered G-OHTL. The company provides an executive charter and engineering and management service from Brooklands Aerodrome, Weybridge.**
*Air Hanson*

***Left:***
**A PLM JetRanger transporting concrete for a hill loch dam.** *PLM*

***Right:***
**McAlpine Helicopters' fleet includes three Aerospatiale types, the SA.365 Dauphin II, the AS.350B Ecureil, and the one illustrated, G-MCAH, an AS.355F Ecureil II.** *McAlpine Helicopters*

***Below:***
**Sykes Aviation of Hurn Airport, Bournemouth is now the only UK operator of the Wessex 60. Seven ex-Bristow machines are in the fleet.** *Peter R. March*

***Bottom:***
**British Caledonian Helicopters' Sikorsky S-61N, G-BIMU, used on the Gatwick Heathrow Airlink. It took over the service at the beginning of 1985 from G-LINK, which had been used since the route's inception in 1978.** *British Caledonian*

**Westland Helicopters Ltd**
Yeovil, Somerset BA20 2YB

*COMMERCIAL HELICOPTER OPERATORS*

**Aeromega Ltd** (1979)
Head Office: Stapleford Aerodrome, nr Romford, Essex RM4 1RL
Fleet: One Bell 206L-1 LongRanger II, one Bell 206B JetRanger, one Piper PA-34-200T Seneca II

**Agricopters Ltd**
Head Office: Martins Lane, Chilbolton, Stockbridge, Hampshire SO20 6BL
Fleet: Four Hiller 12E, two Hiller 12EH Soloys, two Bell 47G-3Bs

**Air Cumbria**
Head Office: Carlisle Airport, Carlisle, Cumbria CA6 4NW
Fleet: One Aerospatiale SA.341G Gazelle, one Cessna 310I

**Air Hanson Ltd**(1962)
Head Office: Brooklands Aerodrome, Weybridge, Surrey KT13 0SH
Fleet: Two Sikorsky S-76As, one Bell 206B JetRanger II, one Agusta-Bell 206B JetRanger II, two Bell 206L LongRangers

**Bond Helicopters Ltd** (1961)
Head Office: Bourn Airfield, Cambridge CB3 7TQ
Fleet: Six Aerospatiale AS.332L Super Pumas, six Aerospatiale SA.365C Dauphins, six Aerospatiale SA.365N Dauphin IIs, nine Bölkow Bö105D, one Agusta A109A, six Sikorsky S-76As. On order four Aerospatiale As.332L Super Pumas, two Aerospatiale SA.365N Dauphin IIs

**Bristow Helicopters Ltd (UH)** (1951)
Head Office: Redhill Aerodrome, Redhill, Surrey RH1 5JZ
Fleet: 22 Aerospatiale AS.332L Tigers, 21 Sikorsky S-61Ns, nine Sikorsky S-76As, 11 Bell 212s, four Agusta-Bell 206A/B JetRangers, one Bell 206A/B JetRanger, one Bölkow Bö.105D, seven Westland 47G-4As, one Bell 47G-2, plus other helicopters frequently registered overseas

**British Airways Helicopters Ltd (BA)**(1964)
Head Office: London (Gatwick) Airport South, Horley, Surrey RH6 0NP
Fleet: Four Boeing-Vertol BV.234 Chinooks, 17 Sikorsky S-61Ns, four Sikorsky S-76s, three Aerospatiale AS.332L Super Pumas, three Westland 30-100s

**British Caledonian Helicopters Ltd** (1979)
Head Office: Farburn Terrace, Aberdeen Airport East, Dyce, Aberdeen, Scotland AB2 0DT
Fleet: Three Bell 214STs, five Sikorsky S-61Ns, one Sikorsky S-76A

**Burnthills Highland Helicopter Service Ltd** (1974)
Head Office: Hangar 12, Glasgow Airport, Scotland PA3 2TF
Fleet: One Agusta-Bell 206B JetRanger, one Bell 206B JetRanger III

**CB Helicopters Ltd** (1978)
Head Office: Westland Heliport, Lombard Road, London SW11 3RE
Fleet: Two Bell 206B JetRanger IIs, one Agusta-Bell 206B JetRanger II

**Cabair Air Taxis Ltd** (1968)
Head Office: Elstree Aerodrome, Boreham Wood, Hertfordshire WD6 3AW
Fleet: One Aerospatiale AS.350B Ecureuil, one Bell 206B JetRanger, one Agusta-Bell 206B JetRanger III

**Castle Air Charter Ltd**
Head Office: Trebrown, Liskeard, Cornwall PL14 3PX
Fleet: One Agusta A109A, one Bell 206L-1 LongRanger, two Agusta-Bell 206B JetRanger IIIs.

**Colt Executive Aviation** (1978)
Head Office: Staverton Airport, Cheltenham, Gloucestershire GL5 6SR
Fleet: Two Aerospatiale AS.350B Ecureuils

**Compass Helicopters** (1982)
Head Office: Main Terminal, Bristol Airport, Bristol BS19 3DY
Fleet: Two Agusta-Bell 206B JetRangers

**Dollar Helicopters** (1978)
Head Office: Hangar No 6, Coventry Airport, Baginton, Coventry CV8 3AZ
Fleet: One Aerospatiale SA.350B Ecureuil, three Aerospatiale SA.315B Lamas, six Agusta-Bell 206B JetRangers, five Bell 206B JetRangers, three Agusta-Bell 47G-3B1s, one Bell 47G-5, five Westland-Bell 47G-3B1s, one Piper PA-23-250 Aztec

**Gleneagle Helicopters** (1981)
Head Office: Edinburgh Airport, Edinburgh EH12 0AL, Scotland
Fleet: One MBB Bö.105D, one Aerospatiale AS.350B Ecureuil, two Agusta-Bell 206A JetRangers, one Enstrom 280C Shark, one Westland 47G-3B1

**Heli Leeds Ltd** (1972)
Head Office: Leeds Heliport, York Road, Whinmoor, Leeds LS14 3AD
Fleet: One Agusta-Bell 206B JetRanger, three Bell 206A/B JetRangers

**Helicopter Hire Ltd** (1969)
Head Office: Aviation Way, Municipal Airport, Southend-on-Sea, Essex SS2 6UN
Fleet: One Westland W30, two Bell 206B JetRangers, two Bell 47G-5A, one Bell 47G-5, one Bell 47G-4A, one MBB Bö.150D, two Westland-Bell 47G-3B1s

**McAlpine Helicopters Ltd** (1978)
Head Office: Swallowfield Way, Hayes, Middlesex UB3 1SP
Fleet: Two Aerospatiale AS.350B Ecureuils, four Aerospatiale AS.355F Ecureuil IIs, two Aerospatiale SA.365N Dauphin IIs, one Westland SA.341G Gazelle

**Alan Mann Helicopters Ltd** (1968)
Head Office: Fairoaks Airport, Chobham, Surrey GU24 8HX
Fleet: Three Agusta A109As, one Agusta-Bell 206B JetRanger III, one Aerospatiale AS.350B Ecureuil, one Bell 206B JetRanger III, one Agusta-Bell 47G-5A

**PLM Helicopters Ltd** (1975)
Head Office: Dalcross Airport, Inverness, Scotland
Fleet; Two Bell 206L LongRangers, one Agusta-Bell 206B JetRanger II

**Red Rose Helicopters** (1983)
Head Office: Hangar 42, Blackpool Airport, Blackpool FY4 2QS
Fleet: One Bell 206B JetRanger, one Agusta-Bell 206B JetRanger III

**Skyhook Ltd** (1981)
Head Office: Vanguard House, Western Avenue, Greenford, Middlesex UB6 8DW
Fleet: One Agusta-Bell 206B JetRanger III

**Skyline Helicopters Ltd** (1983)
Head Office: Wycombe Air Park, High Wycombe, Buckinghamshire SL7 3DR
Fleet: Two Agusta-Bell 206B JetRangers, one Enstrom F28C, one Hughes 369HM, one Robinson R22

**Southernair** (1970)
Head Office: The Flight Centre, Shoreham Airport, Sussex BN4 5FF
Fleet: One Agusta-Bell 206A JetRanger, one Agusta-Bell 206B JetRanger II, three Enstrom 280C-UK2 Sharks, one Enstrom 280C Shark, one Enstrom 28C-UK2, two Hughes 500Es, one Hughes 369D

**Star Aviation Ltd** (1974)
Head Office: Westland Heliport, London SW1
Fleet: Three Bell 206B JetRanger IIs

**Sykes Aviation Ltd**
Head Office: Bournemouth Hurn Airport, Christchurch, Dorset BH23 6DB
Fleet: Seven Westland Wessex 60s

**Veritair Ltd** (1982)
Head Office: Cardiff-Wales Airport, Cardiff, Glamorgan CF6 9BD
Fleet: Two Agusta-Bell 206B JetRanger IIs

***Below:***
**Dollar Helicopters utilises a large fleet of a variety of helicopters, including this Bell 47, on charter, survey and other aerial activities.** *Dollar Helicopters*

***Left:***
**Gleneagle Helicopters of Edinburgh operates the Aerospatiale AS.350B Ecureil.** *Gleneagle Helicopters*

***Below:***
**Helicopter Hire's Westland 30, G-KATE, fitted with its Heli-Tele system.** *Helicopter-Hire*

***Bottom:***
**Robinson R22 G-BLDX of Skyline Helicopters, used for charter and training.** *Skyline Helicopters*

*Right:*
**A Hughes 500E of Southernair at Shoreham Airport.** *Southernair*

*Below:*
**Bell 206B JetRanger, G-BAZN, approaching its Thames base. It is painted in the colours of DHL and flies six times daily with courier packages between Central London and Heathrow Airport.** *DHL*

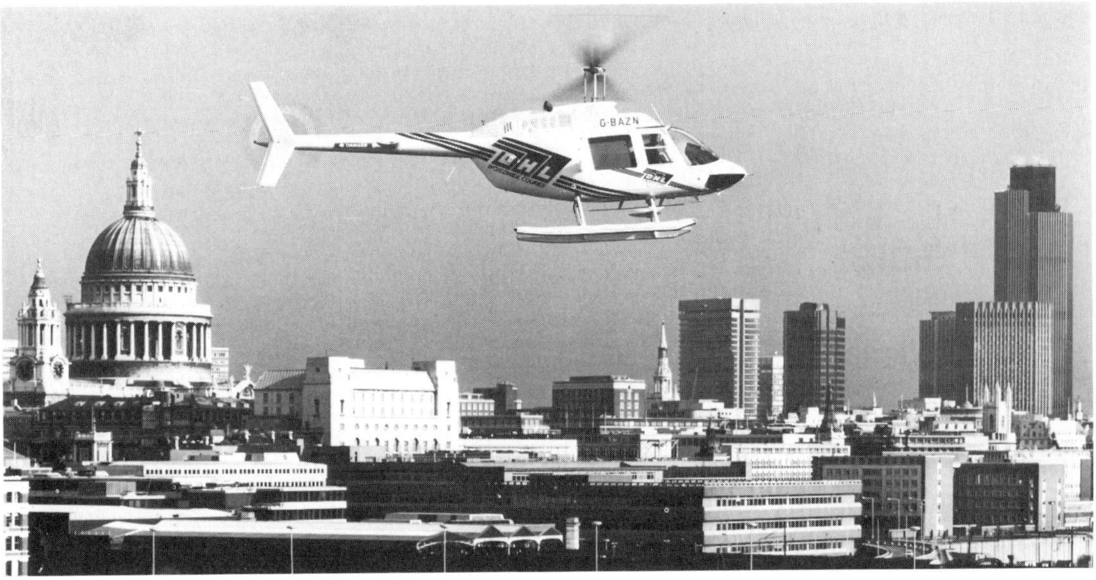


Dollar Helicopters
Dollar Air Services Limited

CB
HELICOPTERS

British
airways
helicopters

British Caledonian
Helicopters

RRH
RED ROSE HELICOPTERS

SOUTHERN
AIR

BHAB
BRITISH HELICOPTER ADVISORY BOARD

BRISTOW HELICOPTERS
LIMITED

STAR AVIATION

VERITAIR
Limited

# British Aircraft Markings

**British Aircraft Markings—**
Civil aircraft markings fulfil two major functions, legal and visual. Each aircraft has a unique registration number allocated by the Civil Aviation Authority, complying with the statutory requirements of the International Civil Aviation Organisation (ICAO), and a paint scheme whereby the operator tries, through colour, insignia and logo-style, to create an image that is both easily identifiable and memorable.

As aircraft technology advances and shapes become ever more futuristic, so visual airline images change, and what is fashionable today will certainly not be so in 10 years time. For an international airline which often acts as an ambassador for the country of its origin, awareness of changing fashions is very important, in order to ensure that its presentation does not become dated. The same applies of course to all airlines right down to small, one-aircraft taxi companies, which have to face strong competition for business. Given a choice, the traveller will always go for the airline with modern aircraft, an attractive image and a good in-flight service. It is on these criteria that airline reputations are built.

**Nationality Marks**
Nationality markings for the identification of aircraft were first suggested at an international convention in Paris in 1910 and, although this principle was adopted in a draft Air Navigation Bill in the UK in 1911, no such system materialised anywhere in the world until after the end of World War 1. In the UK the first Air Navigation Regulations were issued on 20 April 1919, preceding by three months international regulations which came into force on 22 July, following the International Air Navigation Convention held in Paris. In the interim period, two systems prevailed in the UK. Aircraft already bearing RAF serials could retain these as civil markings, subject to the obliteration of the service roundels, and newly-built aircraft were registered in a new sequence starting at K-100. Both systems continued until 22 July, when British aircraft began to be painted with the internationally-agreed four letter registration groups prefixed by the nationality letter 'G-'. The first such series commenced at G-EAAA.

All letters had to be painted in black on white as large as possible on both sides of the fuselage, as well as on the top surface and underside of the wings. In addition, the nationality letter had to be applied to both sides of the tailplane. Gradually, however, many of these rules were forgotten and the present standards, adopted by ICAO on 8 February 1949, call for registrations on the lower surface only of wings, on either side of the rear fuselage, or on the upper half of the vertical tail surface. They must also be clean, clearly visible and identifiable.

In 1928 Britain was allotted a new sequence starting with G-AAAA, the earlier series ending at G-EBZZ. To avoid confusion with the letter 'O', and in deference to the so-called three letter 'Q' codes (ie QNH and QFE), the letter'Q' has not been used since G-EBTQ, a Fokker F.VIIa, in 1927. Lighter-than-air craft were given the batch G-FAAA to G-FZZZ between 1920 and 1928, but only very few were thus issued and did not get beyond G-FAAZ. Airships and balloons are now registered in the main sequence, but microlights, a recent and fast-growing phenomena, are being registered exclusively, for the time being at least, in the G-M--- series. Sequential registrations have now reached G-BLZ- (as at June 1985), the G-A--- series being completed in July 1972.

UK registrations are administered by what is now the Civil Aviation Authority.

**G-UESS Who?**
For almost half a century, the UK registration authorities maintained an orderly sequential register with only isolated instances of re-allocations or special registrations. This began to change on 6 May 1968 with the allocation of G-BSST for Concorde, followed in 1971 by G-VSTO for a Harrier jump-jet. By 1976 there were still only 13 such registrations but since then, under strong pressure from commercial and publicity interests, the floodgates have opened, and there has been a steadily accelerating rate of personalised non-sequential registrations which are currently carried on more than 1,000 fixed-wing aircraft, helicopters, airships and balloons. The situation became so chaotic that a decision was made in 1978 by the CAA to re-establish some sense of order. This basically meant that no existing aircraft with UK marks is to be re-registered, and it is also not intended to re-allocate marks which have already been used. Personalised registrations, however, will continue to be issued subject to the above stipulations.

**Class 'B' Markings**
Britain also uses a system of provisional

***Above:***
**Two photos illustrating early registration sequences — Handley Page HP.42 G-AAGX, and The Instone Air Line de Havilland DH.34A G-EBBV.**
*British Airways*

G-14-1980
AERO SERVICES

G-FOYL
AirFoyle
SKYPAK

Turbine
G-OTVS

registrations for use in UK airspace only, the so-called Class 'B' marks, which may be carried on test, experimental, demonstration or delivery flights prior to the aircraft receiving a full Certificate of Airworthiness. First introduced in 1929, this initially consisted of an individual letter allocated to each manufacturer, followed by the aircraft number separated by a hyphen. For example, Armstrong Whitworth aircraft were registered A-1, A-2 et seq, Blackburn Aircraft B-1, B-2, and so forth. This was modified slightly during World War 2, using the same manufacturer's letter, but followed by zero and three digits, so that the serial looked more like a genuine civil or military registration, thus drawing less attention to the possible experimental nature of the aircraft. From 1948 all aircraft were prefixed by the national letter 'G', followed by a hyphen and a number which identified the manufacturer. A further number related to the individual aircraft identity; eg G-31-12, where 31 indicated Scottish Aviation and 12 the aircraft.

Allocations now in use are much reduced from almost 50 in 1948, and include G-1 Rolls-Royce, G-5 British Aerospace Hatfield, G-9 BAe Kingston, G-11 BAe Manchester, G-14 Shorts, G-16 BAe Weybridge, G-17 Westland, G-27 BAe Warton, G-31 BAe Prestwick, G-51 Pilatus Britten-Norman, G-53 NDN Aircraft, G-56 Edgley Aircraft and G-57 Airship Industries.

## Colour Schemes

On 4 December 1984 British Airways revealed a new image, the first major facelift in more than a decade, replacing the scheme designed by leading British consultants Negus & Negus and carried since September 1973. Designed by Landor Associates of California in association with Chester Jones, a British design company, the livery is intended to produce a strong new identity for the national flag-carrier in time for privatisation. Three colours are used, rich midnight blue for the lower fuselage, brilliant red for the new 'speedwing' superimposed on the blue and running most of the fuselage length, and pearl grey which replaces white on the upper fuselage and lower tail fin. The new 'speedwing' replaces the famous 'speedbird' motif first introduced by Imperial Airways in 1932. New style British Airways titles are traditionally placed, and aligned with the speedwing below. The tail fin retains the quarter Union Jack of the previous livery, but the top section is now midnight blue, promoting the British Airways coat of arms in pearl grey. The first aircraft to be painted up was Boeing 747-136 G-AWNO. The new image is not confined to the fleet exteriors, and will include redesigned aircraft interiors in shades of grey and midnight blue with red and silver highlights, and will also be applied to all other aspects of the airline's activities from sales and ticket offices, ground handling equipment and vehicles, to company stationery and advertising promotions. The new corporate identity will set British Airways back by some £2million, and will take two years to complete.

Another new livery has recently been introduced by British Air Ferries, using to good effect the national colours of red, white and blue. These are also being used by Air UK, Britannia Airways and Metropolitan Airways. Regional symbols are often used to indicate an airline's origins and include the Rampant Lion of Scotland by British Caledonian set in gold into a blue tail; the Red Dragon of Wales by Airways International Cymru; and Manx Airlines, not surprisingly, features the Three Legs of Man, also known as the Trinacria, in its striking red and green livery. Air Europe uses typical holiday colours of yellow, orange and blue, symbolising golden sun, sandy beaches and warm blue skies and water, as does Orion Airways.

Monarch Airlines aircraft are painted with broad gold and black cheatlines, with the 'M' stylised into a crown in black on a white tail. Other airlines use the initials of their name to good effect on the aircraft fin, the most prominent being British Midland (BMA), British Air Ferries (BAF), and British Island Airways (BIA).

***Top left:***
**This Shorts Skyvan with 'Class B' markings G-14-1980 was crew-training at Hurn at the end of 1984 prior to delivery to Barbados.** *Andrew March*

***Centre and Below:***
**Personalised, out-of-sequence registrations are often instantly recognisable, like Air Foyle's Piper Aztec G-FOYL and Islander G-OTVS of Television South.**

# Britain's Airports

Prior to World War 2, all civil aerodromes in the United Kingdom, with the exception of Croydon, Heston and Lympne, were owned and operated by municipal authorities or private companies. All of these were either requisitioned or taken over by the government during the war years when all civil flying was banned. When the war ended Britain had about 700 airfields and the new Labour government, in its 1945 White Paper on 'British Air Services', decided that all airfields required for scheduled air services were to be acquired and managed by the Ministry of Civil Aviation. It was not until 1947 however, that the Ministry drew up a list of 44 airfields, excluding the London airports, which were to be acquired to form a national airport plan. The airports taken over made substantial losses and it was not long before the Ministry began 'courting' the local authorities to take them off its hands.

Beginning with the transfer of Ringway Airport to Manchester Corporation in 1950, the growing municipalisation of airports culminated in the 1961 White Paper on 'Civil Aerodromes and Air Navigational Services' which stated that the government would divest itself of the remaining airports still under its ownership. No state subsidies would be provided unless the airports were considered indispensable to the national air transport system. The White Paper also proposed that the four major international airports should be taken over by a public corporation, and on 1 April 1966 the British Airports Authority came into being as the owner of Heathrow, Gatwick, Stansted, and Prestwick in Scotland.

Thus by the end of the 1960s ownership of airports in the British Isles was in the hands of a mixture of central government, individual city councils, consortia of local authorities, private companies, and even aircraft manufacturers, and development consequently progressed in a rather haphazard manner. There was no lack of investment as airports tried to attract traffic by providing better facilities than their close neighbours, but such rivalries failed to stem the growing concentration of traffic at the London area airports.

During the last decade, calls for greater rationalisation and co-ordination of regional airport development became increasingly vociferous, and this resulted in a White Paper on 'Airports Policy' which was published in February 1978. It divided the airports in England and Wales into four categories:

A: Gateway International Airports
B: Regional Airports
C: Local Airports
D: General Aviation Aerodromes.

Belfast and the Scottish airports were for some obscure reason not so categorised.

Since the publication of the White Paper there has been a trend towards concentration, and the growth of international passengers from the regions has been particularly marked at Manchester, Birmingham and Glasgow airports. Success stories however are few, and uncertainties continue to hang over the future of some of the smaller airports, even though the government has pledged capital projects worth £31million in 1985/86 to the development of regional airports. With the present government's intention to abolish the metropolitan county authorities, the future ownership of some local authority airports is under review.

**The British Airports Authority**

BAA owns and manages seven major airports in the UK: Heathrow, Gatwick and Stansted serving London, and Glasgow, Edinburgh, Prestwick and Aberdeen in Scotland. These seven airports handle 75% of the nation's air passenger traffic and 85% of its air cargo. In 1984 49.9 million passengers passed through the seven BAA airports, an increase of 10% over the previous year. It is by far the largest airport operator in the UK and was established as a national enterprise in 1966 initially taking over Heathrow, Gatwick, Stansted and Prestwick; Edinburgh was acquired on 1 April 1971; Aberdeen on 1 January 1975; and finally Glasgow on 1 April 1975.

British Airports provides a wide range of services and facilities, including planning, developing and maintaining airport buildings, runways and roads; and providing fire services, apron control, terminal management, security, passenger services and information systems, restaurants, shops and car parks. Also provided are consultancy and project management services for overseas airport authorities and aviation authorities through its subsidiary, British Airports International Ltd.

The major part of its income — more than half — comes from aircraft landing and parking charges paid by the airlines. But this revenue alone would not cover the cost of providing the necessary airport facilities and loss on traffic operations is turned into profit by income from

# Major UK Airports

***Above:***
**A British Airways Tristar at Heathrow Airport.** *BAA*

***Below:***
**Heathrow central area from the air.** *CAA*

commercial enterprises such as duty free sales, catering, car parking and other activities undertaken by concessionaires appointed by BAA. As a result of the rapid growth of air travel, much of its revenue has been invested in new and expanded facilities. In its 17 years of existence capital expenditure approached £400million, but this sum looks insignificant when compared with the proposed £780million programme for the five years from 1983 onwards.

**The London Airports System**

The three London airports of Heathrow, Gatwick and Stansted form the most successful international airport system in the world offering a wide range of scheduled services with well over 2,000 departures a week to 230 international destinations.

**Heathrow** is the world's busiest international airport and the busiest of any kind outside the USA. In 1984 it handled over 29 million passengers for the first time, and is used by over 70 airlines. It is the UK's leading port in value of visible trade; in 1983 this amounted to £16.6 billion.

The history of the area now known as Heathrow Airport-London began in 1929 when Fairey Aviation built its Great West Aerodrome. It was used mainly for experimental flying until being requisitioned by the Air Ministry in 1944. The war ended before the airfield could be put to military use and a decision was taken to develop the site into a major civil airport for London.

The three passenger terminals are grouped around the control tower and the central bus and underground stations. Terminal 1 handles almost 11 million passengers a year which would make it Europe's sixth busiest airport in its own right. It handles all Heathrow's domestic traffic, a large volume of British Airways' European flights and routes served by Aer Lingus. Terminal 2 is used by 29 non-UK airlines operating services to Europe, and handles six million passengers a year. The Eurolounge connects Terminals 1 and 2 and is used primarily as a gateroom for high-density European routes providing a waiting area for up to 700 passengers. Terminal 3 has 35 long-haul and intercontinental airlines and handles about 10 million passengers per annum.

A fourth terminal, costing £160 million is due to be opened in October 1985. This new terminal, sited away from the central area, will have its own underground station and road access to the M25 motorway. It has been designed for a mix of short and long haul traffic and, together with associated taxiways, aircraft stands and car parks, will raise Heathrow's capacity from the present 30 million passengers per annum to 38 million per annum. When the fourth terminal opens, air transport movements at Heathrow will be limited to 275,000 per annum for environmental reasons.

The building of a fifth terminal by the mid-1990s on the site of the present Perry Oaks Sewage Works, to raise the annual passenger capacity to 53 million, has been shelved for the time being.

**Gatwick** is situated 45km south of London on the M23. Apart from good road connections, it is also linked to the capital by a frequent non-stop rail express service which was inaugurated on 14 May 1984. A helicopter link service provides 10 daily flights in each direction between Gatwick and Heathrow airports for transit passengers.

In 1984 Gatwick's annual passenger total was just under 14 million, an above average improvement of 11.8% over 1983. In terms of the number of passengers handled on international flights, Gatwick with around 12 million is currently the fourth busiest airport in the world behind Heathrow (22.4 million), New York-JFK (14.6 million) and Frankfurt (12.2 million). Freight figures for 1984 amounted to 143,000 tonnes. It has one main terminal and its single 3,098m runway is capable of handling all civil aircraft currently in service. To accommodate future air traffic needs, a second passenger terminal, known as the North Terminal, is now under construction. It will be coming into use in 1987 and will be able to accommodate a further nine million passengers per annum, increasing the airport's capacity to 25 million passengers by the early 1990s. Though often considered in the past, BAA has now ruled out provision of a second runway at Gatwick.

There has been an airfield at Gatwick since 1930 when a private aerodrome was opened near Lowfield Heath, about a mile south of the present terminal building. After the war Gatwick was chosen by the government as the site for London's second airport, primarily to take diversions from Heathrow and serve as a base for independent airlines. The old airport was closed in 1956, and a completely new airport, built at a cost of £7.8 million, was opened by HM The Queen on 9 June 1958. Further development work in 1962, 1968 and throughout the following years provided for improved passenger facilities, a runway extension, car parking and ancillary services. On 21 April 1983 a satellite terminal, linked by Europe's first fully-automatic rapid transport system was opened. This was the last stage of a major development programme to expand the capacity of the existing terminal to 16 million passengers per annum.

**Stansted** has figured prominently in the 20-year debate about the future of the London

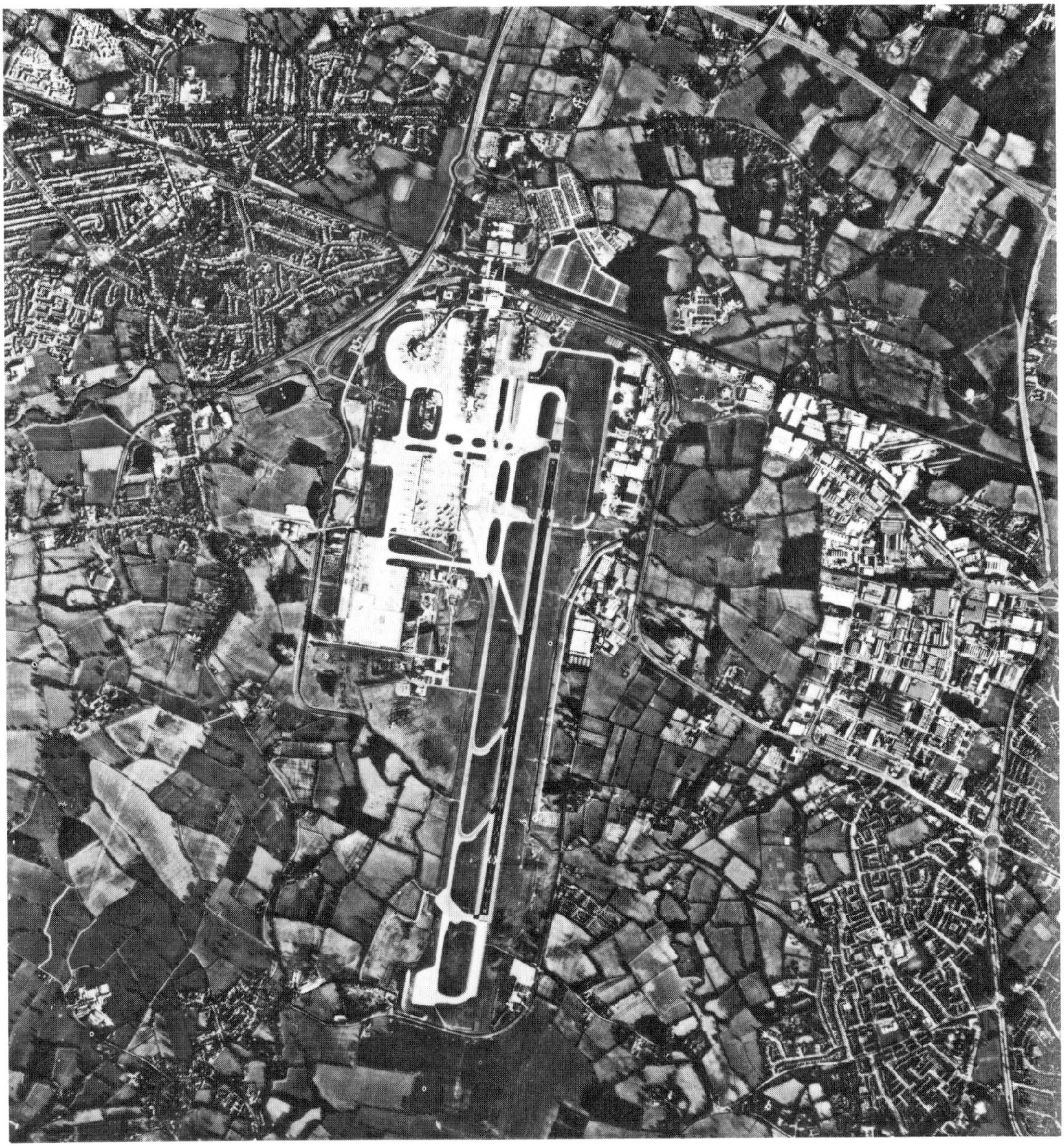

*Above:*
**An aerial view of Gatwick Airport.** *BAA*

*Left:*
**A big day for Stansted Airport in 1983 — the visit of the United States space shuttle *Enterprise*.**
*Malcolm Barber*

airport system, but its history goes back beyond that. It began in 1942 with the construction of a United States Air Force base and was taken over by the UK government after the end of the war.

Stansted is situated 56km to the northeast of London and has good road access (via the M11 and M25 motorways) from London, the Southeast, East Anglia and the Midlands. In addition to holiday charter flights it has an expanding network of scheduled services to domestic destinations such as Edinburgh, Humberside and Jersey, and internationally to Amsterdam, Brussels and Paris. In 1984 Stansted handled 527,100 passengers, a growth rate of 54% on 1983, due mainly to an increase in European scheduled traffic and a 50% jump in Scandinavian charter activity.

In 1979 the government invited BAA to prepare proposals for the development of Stansted Airport to be examined by public inquiry. In September 1981 the Authority put forward plans for a new terminal capable of handling 15 million passengers a year, with its own link to the M11 motorway and a railway station providing a convenient connection with London's St Pancras. The new facilities have been costed at £400million at 1981 prices. BAA's proposals have been adopted in a report published in December 1984. The report, the culmination of a 258-day public enquiry, calls for an immediate expansion of Stansted as a single-runway airport capable of handling 15 million passengers annually, rising to 25 million by the mid-1990s. Mr Graham Eyre QC, the government-appointed inspector, warns in his report that the present and planned capacity of the London Airports System (in which he includes Luton) will not be sufficient to meet demands much beyond the end of this decade. It is expected that the four airports will be handling 47 million passengers in 1985, rising to 61 million by 1990, 75 million by 1995, and 89 million by the year 2000. In June 1985 Transport Secretary Nicholas Ridley ended a quarter century of indecision by approving limited development of Stansted for up to 8m passengers by 1990.

Much of the future development and administration of the London Airports System will be dependent on a successful privatisation of the British Airports Authority in line with present government policy.

### London STOLport

The government has given the go-ahead for the construction of a STOLport in London's Dockland to allow passenger aircraft to land in the centre of the capital. The £10million STOLport will be sited between the Royal Albert and King George V docks on the north bank of the Thames, and will include new terminal and hangar facilities, and a new 750m runway. The proposal was originally put forward by the construction group **Mowlem and Company**, and by **Brymon Airways**, which hopes to begin services with its DHC-7s by summer 1986. In the first year services are planned to Birmingham, Newquay, Plymouth, Amsterdam, Brussels and Paris, with Aberdeen, Belfast, Frankfurt, Glasgow, Guernsey, Jersey and Manchester to follow later.

### BAA Scottish Airports

The four BAA Scottish airports constitute a major system of regional airports, handling 6.3 million passengers in 1984, half a million more than in 1983. Glasgow and Edinburgh serve Scotland's two major cities for domestic and short-haul international flights, while Prestwick, on the Ayrshire coast, is Scotland's transatlantic gateway. Aberdeen has been the fastest growing airport in the country over the past decade because of its strategic position for North Sea oil operations. Investment has been minimal compared to that for the London airports with a total of £50million expended so far and another £40million allocated in the mid-1980s.

**Glasgow (Abbotsinch)** is the fourth busiest airport in the UK and an important passenger and cargo terminal for the industrial and commercial west of Scotland. From there, services radiate to other parts of the UK, including the Scottish Highlands and Islands, and there are routes to European destinations such as Amsterdam, Dusseldorf, Geneva, Reykjavik, Stavanger, Dublin, Copenhagen and Paris. Glasgow's busiest route is to London, carrying some 900,000 passengers a year, which is about one third of the total of 2.75 million handled in 1984.

The present airport at Abbotsinch on the south bank of the Clyde was originally constructed in 1932 as a military aerodrome and did not come into civil use until 31 October 1963 when it was handed over to the Ministry of Civil Aviation. Although Renfrew had served as Glasgow's airport since the start of Scottish air services in 1933, it could not be sufficiently developed to meet the city's future needs, and in November 1960 Abbotsinch was designated as a replacement, and became the new Glasgow Airport on 2 May 1966. Ownership passed from Glasgow Corporation to BAA on 1 April 1975.

**Prestwick** is Scotland's intercontinental gateway. Situated on the Ayrshire coast, it has an exceptionally good weather record, often handling diversions from other UK airports when they are affected by fog or other bad weather.

# THE PROPOSED ROYAL DOCKS STOLPORT

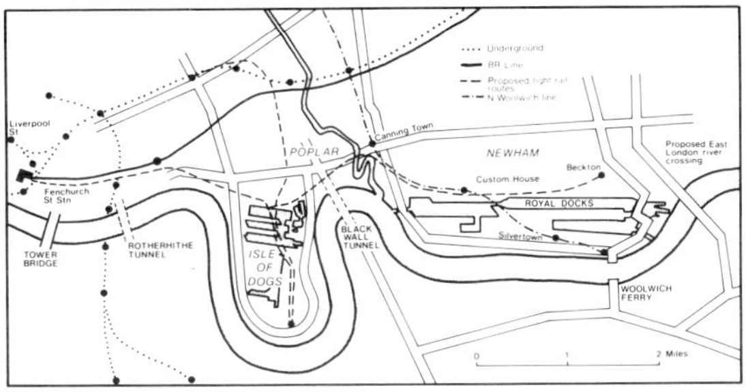


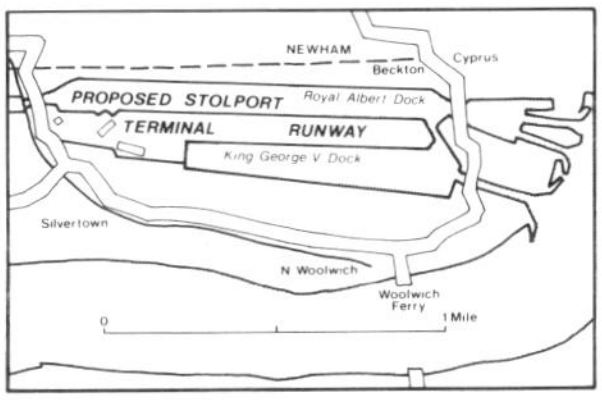


***Below:***
**A de Havilland Canada DHC-7 of Brymon Airways lands in London Docklands to demonstrate the feasibility of Mowlem's proposal to build and operate a STOLport in the Royal Docks.**
*Brymon Airways via Malcolm Ginsberg*

***Bottom:***
**Glasgow Airport from the air.** *BAA*

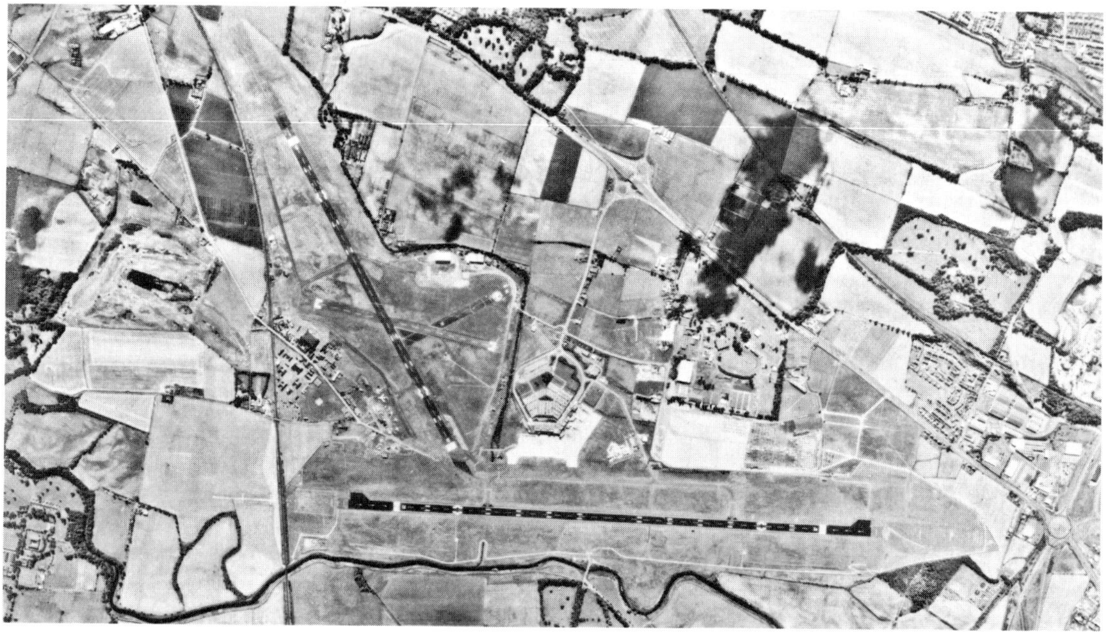

***Top:***
**Prestwick Airport.** *BAA*

***Above:***
**An aerial view of Edinburgh Airport.** *BAA*

Because of its good weather it is also used by airlines for training, test flying by British Aerospace and by Royal Navy helicopters on anti-submarine and rescue duties.

Most of Prestwick's scheduled flights are to and from North America, maintaining the airport's long association with transatlantic flights which began during World War 2, when it was used as a base for ferrying aircraft across the Atlantic. Present scheduled services are those of Northwest Orient to Boston and New York, and Air Canada's services to Halifax and Toronto; Highland Air hopes to provide low-cost flights to New York and Toronto from spring 1985. At home, Air Eçosse provides scheduled flights to Aberdeen and Belfast.

Some 235,000 passengers and 10,300 tonnes of cargo passed through Prestwick in 1984, which is a small decline over the previous year. Determined efforts are being made by BAA and Scottish trade and tourist organisations to improve the fortune of the airport by encouraging direct transatlantic visits to Scotland. Prestwick is also ideally situated to serve any potential development of oil exploration in the Firth of Clyde. In January 1984 it was designated as the site of one of Britain's six proposed freeports.

Prestwick's history goes back to 1935 when it was set up as a grass flying field by the Duke of Hamilton and the late Grp Capt D. McIntyre. It was transferred to the RAF at the beginning of the war and came into the ownership of BAA, via the Ministry of Civil Aviation, in 1966.

**Edinburgh (Turnhouse)** serves the Scottish capital and is located 11km west of the city. It began as a grass airfield in 1915 when it was used by the Royal Flying Corps for training purposes. It remained a military field between the wars and continued under military control afterwards, before coming under the control of the Ministry of Civil Aviation.

The conditions of transfer to BAA included a government grant covering 75% of the cost of a new runway and terminal complex to meet future traffic. The new 2,560m runway was brought into operation on 7 April 1976, and the terminal on the south side was opened by HM The Queen on 27 May 1977. The terminal has since been improved and enlarged to handle two million passengers. Edinburgh Airport presently deals with some 1.5 million passengers per annum, most of whom travel on domestic services to London. In addition to flights to other parts of Britain, Edinburgh caters for a wide range of holiday charter destinations and has scheduled services to a number of European cities including Amsterdam, Dublin, Dusseldorf, Frankfurt and Paris.

**Aberdeen's** emergence as a major airport has been greatly stimulated by North Sea oil exploration and production. As a result it has been one of Europe's fastest growing airports handling 1.76 million passengers in 1984. Of these, 1.1 million were carried on fixed-wing aircraft, and 660,000 travelled by helicopter, which makes Aberdeen the world's busiest heliport, handling the largest rotary-wing aircraft in current service. Cargo tonnage amounted to 8,500 tonnes, over half of which was transported by helicopters to the oil rigs. In terms of aircraft movements, it is the third busiest airport in Britain.

Aberdeen offers direct scheduled connections to a range of destinations in the UK and Europe, including other 'oil towns' such as Stavanger and Bergen. In addition to domestic and international traffic, inclusive tour services operate during the holiday period to various destinations through Europe. The airport also has a flourishing general aviation business, again mainly associated with the oil-related industries.

Life for Aberdeen's Airport at Dyce began in March 1931 when Eric Gandar Dower made a landing in his Blackburn Bluebird. During the war the airport was requisitioned and became an important base for fighters and Coastal Command. It was transferred to BAA's Scottish Airports group from the Civil Aviation Authority in January 1975.

The other major airport owner in Scotland is the **Civil Aviation Authority** which operates Inverness, Kirkwall, Wick, Stornoway, Benbecula, Tiree and Islay.

**Manchester International Airport (Ringway)**

Manchester International Airport is the UK's largest and busiest airport outside London. At the heart of the industrial region, its catchment area extends from the Midlands and throughout the whole of the North of England serving 20 million people and almost half the entire country's manufacturing industry. It is owned jointly by the City of Manchester and Greater Manchester County Council, which administer it through the Airport Authority Joint Committee.

Since its designation by the government in 1978 as Britain's only Category A International Gateway airport outside London, it has become one of the fastest growing airports in Europe. In 1984 more than six million passengers passed through the airport and this figure is confidently expected to reach nine million by the end of the decade. As in the past, most of the growth will come from holiday traffic, but it is hoped that a more liberal licensing approach will bring more international services to the airport. MIA Authority is investing over £100million in the future development of the airport. Elements of this

***Above:***
**The striking ziggurat shape of the control tower at Aberdeen Airport from the outside, and the nerve centre from the inside.** *CAA*

programme (such as the runway extension to 3,050m) have already been completed, and additional parking stands are under construction. Planned for completion by 1990 are a new cargo centre, terminal extensions, a new domestic passenger module, relocation of existing general aviation facilities on the south side of the main runway and various peripherial airport services. One of the most significant parts of the development will be the provision of a rail link which will serve the passenger terminals from the nearby InterCity station.

Over 40 airlines now use the airport regularly, and scheduled services extend as far as Australia and the Far East. There are also scheduled cargo flights to Amsterdam, Dublin, Frankfurt, Helsinki and Zurich.

Work on the present airport site at Ringway began in 1935, and the airport was officially opened by Sir Kingsley Wood, the Secretary of State for Air, on 25 June 1938.

## THE REGIONAL AIRPORTS

### The New Birmingham International Airport (Elmdon)

Birmingham International was opened to traffic on 4 April 1984 with the start of regular services from the city's new air terminal and was formally inaugurated by HM The Queen on 30 May.

The new airport has been hailed as the runway to recovery for the ailing fortunes of the West Midlands region. Birmingham's position in the centre of England's motorway system, an InterCity railway station and the successful National Exhibition Centre at its doorstep, together with the prospects of lower air fares in Europe, should combine to make Birmingham International into a major hub for links with its EEC trading partners. The new £62million three-storey passenger terminal and associated handling facilities have been designed to cater for up to three million passengers and 33,000 aircraft movements a year by 1990. The terminal is linked to its BR station and the NEC by MAGLEV — a new revolutionary robot passenger shuttle which floats on a magnetic cushion 10m above ground level along an elevated track. The journey takes just 90sec.

The presently small amount of cargo handled will be boosted by new freight handling facilities — the first phase, which includes a 6,000sq m bonded freight centre, is now under construction. The prospect of an inland freeport at Elmdon will also attract more business.

Both British Airways and British Midlands have a long association with Birmingham, and their services, together with those of Brymon Airways and Dan-Air/Metropolitan Airways, currently connect the city with several continental and domestic points. The city's own airline, Birmingham Executive Airways, provides regular air services to Copenhagen, Stockholm, Geneva, Stuttgart, Milan and Zurich. Among the foreign airlines flying regular services are Wardair, Air India, Cyprus Airways, Aer Lingus, JAT and NLM. Inclusive tours and charter flights are also growing. Passengers handled in 1984 amounted to 1,752,746, an increase of 8.6% over the previous year.

The history of the airport dates back to May 1939 when Railway Air Services transferred its operations from Castle Bromwich to the new municipal airport at Elmdon. A few months later the airport was requisitioned and, after several changes of ownership, passed to the West Midlands County Council on 1 April 1974.

**Local Airports**

| *Airport* | *Owner/Operator* | *Passenger traffic 1984* | *Scheduled services destinations* |
|---|---|---|---|
| Belfast (Harbour) | Shorts | 177,882 | Blackpool, Edinburgh, Glasgow, Isle of Man |
| Blackpool (Squires Gate) | Blackpool Corporation | 157,686 | Belfast, Londonderry, Isle of Man, Jersey, Liverpool, London (Gatwick), Tees-side |
| Bournemouth (Hurn) | Airport Management Committee | 155,674 | Alderney, Birmingham, Cardiff, London (Gatwick), Cherbourg |
| Bristol (Lulsgate) | Bristol City Council | 446,408 | Belfast, Cardiff, Glasgow, Guernsey, Isle of Man, Jersey, Leeds, London (Gatwick), Cork, Dublin, Brussels, Amsterdam |

| *Airport* | *Owner/Operator* | *Passenger traffic 1984* | *Scheduled services destinations* |
|---|---|---|---|
| Carlisle (Crosby) | Carlisle City Council | 22,939 | Aberdeen, Dundee, Isle of Man, London (Heathrow) |
| Coventry (Baginton) | Coventry District Council | 79,635 | Jersey, Paris (Charles de Gaulle) |
| Dundee (Riverside) | Tayside Regional Council | 32,986 | Aberdeen, Carlisle, Isle of Man, London (Heathrow) |
| Exeter (Clyst Honiton) | Exeter Airports Limited | 108,001 | Glasgow, Guernsey, Isles of Scilly, Jersey, London (Gatwick), Plymouth, Southampton |
| Guernsey | States of Guernsey | 631,061 | Most UK cities and Amsterdam, Cherbourg, Dinard |
| Humberside (Kirmington) | Humberside County Council | 142,426 | Aberdeen, Glasgow, Guernsey, Jersey, London (Heathrow and Stansted), Norwich, Tees-side, Amsterdam, Esbjerg |
| Inverness (Dalcross) | Civil Aviation Authority | 159,535 | Aberdeen, Edinburgh, Kirkwall, Stornoway, Wick |
| Isle of Man (Ronaldsway) | Isle of Man Airports Board | 299,807 | Belfast, Blackpool, Bristol, Carlisle, Dundee, Edinburgh, Glasgow, Jersey, Liverpool, Manchester, Tees-side, Cork, Dublin |
| Jersey | States of Jersey | 1,440,283 | Most UK Cities and Amsterdam, Cherbourg, Dinard, Paris |
| Liverpool (Speke) | Merseyside County Council | 228,084 | Aberdeen, Belfast (Harbour), Blackpool, Isle of Man, London (Heathrow and Gatwick), Dublin |
| Newquay (St Mawgan) | Brymon Airways | 30,000 | Plymouth |
| Norwich (Horsham St Faith) | Airport Joint Committee | 192,732 | Aberdeen, Glasgow, Guernsey, Humberside, Jersey, Leeds, London, Tees-side, Amsterdam |
| Plymouth (Roborough) | Brymon Airways | 86,642 | Aberdeen, Exeter, Guernsey, Isles of Scilly, Jersey, London, Newquay, Cork |
| Southampton (Eastleigh) | Southampton Airport Ltd | 299,976 | Alderney, Exeter, Guernsey, Jersey, Amsterdam |
| Southend (Rochford) | British Airports International | 100,781 | Guernsey, Jersey, Billund, Rotterdam, Le Touquet, Ostend |
| Tees-side (Middleton St George) | Tees-side Airport Committee | 296,206 | Aberdeen, Belfast, Blackpool, Glasgow, Guernsey, Humberside, Isle of Man, Jersey, Leeds, London, Norwich, Amsterdam |

***Above:***
**Cargo being loaded on to a Qantas Boeing 747 at Manchester International Airport.** *MIA*

***Right:***
**The revolutionary 'Maglev' transportation system at the new Birmingham International Airport.**
*Birmingham Airport*

***Below:***
**Aerial view of Manchester International, Britain's only Category A airport.** *MIA*

**Newcastle International Airport (Woolsington)**
Newcastle Airport is situated 9km northwest of the city and adjacent to the A696 trunk road to Scotland. It is controlled by the North-East Regional Airport Committee which is made up from representatives of the Tyne & Wear Metropolitan County Council, Durham County Council and Northumberland County Council. Newcastle serves a catchment area of over 1½ million people and the steady and consistent growth pattern in the northeast bodes well for the future.

With the recent completion of an £8million redevelopment of the terminal building, part of the master plan to take the airport into the 1990s, Newcastle International is now fully equipped to handle up to two million passengers a year. The next stage of development will be the construction of a taxiway parallel to the single 2,332m long runway. The latest Marconi-built airfield approach system was installed in 1983 at a cost of £305,000.

Scheduled passenger services from Newcastle now reach Aberdeen, Amsterdam, Belfast, Bergen, Birmingham, Bournemouth, Cardiff, Dublin, Glasgow, Isle of Man, Jersey, London, Manchester, Norwich, Paris and Stavanger. Passengers carried in 1984 totalled 1,250,000.

Newcastle Airport was opened on 26 July 1935 and, with the exception of the war years, remained in the hands of the City Council until the formation of the North-East Regional Airport Committee in April 1963.

**East Midlands Airport (Castle Donington)**
East Midlands Airport, located on the A453 19km southeast of Derby, is operated by a joint committee made up of representatives of the Derby, Leicestershire and Nottinghamshire County Councils, and the Nottingham City Council. It was the first entirely new municipal airport to be built in England since World War 2, and became operational in April 1965 after being completed in 13 months at a capital cost of £1,375,000.

A development plan review for EMA over the next few years and for the longer term up to 1993 expects an increased demand for the use of the airport with passenger figures up to two million and freight to 50,000 tonnes a year. Estimated investment to 1987-88 will be about £15million which includes £8million for a runway extension to 2,895m. In addition, further expansion to the terminal will be inevitable to accommodate the traffic growth forecast.

The major airline based at East Midlands is British Midland Airways whose routes serve Amsterdam, Belfast, Glasgow, Guernsey, Jersey, London-Heathrow and Paris. Air Eçosse serves Aberdeen, Edinburgh and Wick in Scotland, and there are services to London-Gatwick and to the Isle of Man by Brymon Airways and to Brussels by London European Airways. Nevertheless, EMA's traffic is made up principally of holiday charters for which the airport is one of the country's main gateways. One million passengers used the airport in 1984.

Freight also plays an important and growing part in its future. In the year which ended on 31 March 1984 air cargo and air mail traffic grew by 30% to reach 12,500 tonnes. This improvement is due in no small part to the excellent facilities and the presence of Air Bridge Carriers which is the main all-cargo operator based at the airport. The airport is also a distribution centre for the Datapost and Elan overnight delivery services.

**Leeds/Bradford Airport (Yeadon)**
Leeds/Bradford Airport is owned and operated by a consortium of three local authorities — West Yorkshire Metropolitan County Council, Leeds City Council, and the City of Bradford Metropolitan District Council. It is situated on the A568, 13km to the northeast of Leeds and 10km north of Bradford, and was officially opened on 17 October 1931 as the Leeds/Bradford Municipal Aerodrome. After military use during the war the airport was handed over by the Ministry of Aircraft Production to the Ministry of Civil Aviation in 1947. The Leeds/Bradford Airport Joint Committee took over on 8 January 1958.

Leeds/Bradford Airport serves a catchment area which covers most of the Yorkshire and Humberside region with a population of 4.8million. Although this is larger than that of East Midlands and Newcastle, Leeds/Bradford has lost out through the operational limits imposed by its runway length. This is being rectified with a £23million improvement scheme which includes a 640m runway extension, new airfield lighting, navaids, support services, passenger terminal enlargements and a new cargo area. The new facilities, it is hoped, will upgrade Leeds/Bradford to a major regional airport, and it is confidently expected that by 1995 the annual passenger throughput will treble to 1.4million. 1984 figures include 435,575 passengers, the highest ever, and 2,800 tonnes of freight.

Main operators currently using the airport are Air UK, British Midland, Orion, Monarch, Air Europe, Dan-Air, Britannia, B.Cal, Aer Lingus and Air Malta. Scheduled routes serve Aberdeen, Amsterdam, Belfast, Bristol, Cardiff, Dublin, Edinburgh, Glasgow, London, Norwich, Paris and Tees-side.

***Top:***
**Aerial view of Newcastle International Airport, with Concorde on the apron.** *Newcastle Airport*

***Above:***
**A British Midland Airways McDonnell Douglas DC-9 and Orion Airways Boeing 737 share the apron at East Midlands Airport. The airport is home to both airlines.** *Ralph Wright*

***Left:***
**An artist's impression of the Terminal One extension at Leeds/Bradford Airport which is due for completion in summer 1985. A new runway extension was completed in autumn 1984.**
*Leeds/Bradford Airport*

### Belfast International Airport (Aldergrove)

Situated on the A57, 23km northwest of Belfast and 5km east of Lough Neagh, the airport's history goes back to 1918 when an aerodrome was built near Crumlin. When Midland & Scottish Air Ferries began a scheduled service on 1 June 1933 Aldergrove served as the Belfast Airport, but was soon replaced by Newtownards, then Sydenham and, after the war, by Nutts Corner. On 26 September 1963 all services were once again transferred to Aldergrove. The airport is now operated by Northern Ireland Airports Ltd.

£25million of capital works during the last seven years — including improvements to terminal, car parking and other facilities, and upgrading to CAT III standard which allows full operation in virtually any weather — have turned BIA into one of the most modern airports in the British Isles. The airport received a major boost in February 1984 with its designation as the site for the freeport for Northern Ireland.

Flights serve the main domestic centres as well as Amsterdam and Paris, and there is a sizeable charter traffic to Europe and across the North Atlantic. Scheduled airlines include British Airways, Air UK, Dan-Air, British Midland, Air Eçosse, Loganair, Spacegrand, Aer Lingus and NLM. A separate apron and terminal located to the east of the main terminal has been provided for air taxi, business, private and club aircraft. The airport handled 1.4million passengers during 1983.

### Cardiff-Wales Airport (Rhoose)

Cardiff-Wales Airport, located 16km west of the city, is operated by a consortium of the county councils of Mid, South and West Glamorgan. Since its designation as a Category B regional airport in 1978, there has been steady if unspectacular progress. Passengers handled in 1983 amounted to 406,673, an encouraging increase of 13.5% over the previous year but, in spite of a modern cargo depot, freight traffic shows little upward movement. The bulk of the passenger figures are provided by IT and holiday charter flights which now service all the European sunspots and extend to Canada, the Middle East, North Africa and Russia.

Scheduled services link Cardiff with Amsterdam, Belfast, Bournemouth, Bristol, Dublin, Glasgow, Guernsey, Isle of Man, Jersey, Leeds, London, Manchester and Newcastle, flown by Dan-Air and British Caledonian. Air taxi and executive charter services are also available, and based at the airport is the Dragon Engineering Co which produces the Dragon microlight aircraft.

The airport at Rhoose was originally constructed for the RAF and became Cardiff's new civil airport on 1 April 1954. Extensive redevelopments in the 1970s included a new terminal capable of handling up to one million passengers, runway extension to the present 2,134m, and ancillary services.

### Luton International Airport

Classified as a Category A airport, Luton is now one of the busiest airports in Britain, handling almost two million passengers during 1984. Luton's claim to fame has been built on holiday charter traffic for which the airport, just 45km north of London, has become a major gateway. More than 15 British and European airlines fly regularly to the Mediterranean, and for two of them, Britannia Airways and Monarch Airlines, Luton is home. Also based at the airport are London European Airways, operating a scheduled passenger service to Brussels and Amsterdam, and McAlpine Aviation, one of the country's leading executive charter firms.

Luton Airport was officially opened on 16 July 1938 and is operated by the Luton Borough Council.

### *News in Brief*

**Bristol Airport** has been given the go-ahead for a £6.4million development project which will include new arrival and departure facilities.

**Liverpool Airport's** new £4million terminal building is due for completion by March 1986.

---

**British Airports Authority**
Gatwick Airport
Crawley
West Sussex

# Britain's Civil Aviation Authority

**Britain's Civil Aviation Authority—**
Since the start of Air Transport & Travel's first commercial air service from London to Paris on 25 August 1919, the development of civil air transport in the United Kingdom has been regulated by a succession and mixture of government and independent bodies. This never proved totally satisfactory and in 1967 the Government set up a review committee under the chairmanship of Sir Ronald Edwards to inquire into Britain's civil air transport. The Committee's findings were published two years later in its report *British Air Transport in the Seventies* and one of the principal recommendations was the establishment of a Civil Aviation Authority. The new Authority was to bring together various civil aviation functions previously undertaken by the Department of Trade & Industry, the Air Transport Licensing Board and the Air Registration Board — a division of responsibilities which was regarded by the Committee as a source of weakness in the system as a whole.

The Committee envisaged that the Civil Aviation Authority (CAA) would undertake a more positive role in the economic regulation of the industry, and improve co-ordination between the airworthiness and operational aspects of safety as well as between safety and economic control. In general it was envisaged that the CAA would be 'a strong body standing up for British civil aviation and making the citizens of this country aware, through its reports and pronouncements, of the importance of aviation and Britain's part in it'.

Parliament accepted the recommendations of the Edwards Committee, and provision for an Authority was made in the Civil Aviation Act 1971. The Authority was formally set up on 22 December 1971, and assumed its full responsibilities on 1 April 1972.

The CAA's principal responsibilities now include the following:

- National Air Traffic Services — air traffic control and telecommunications, jointly with the Ministry of Defence.
- The economic regulation of the civil aviation industry,including air transport licensing and approval of air fares, and the licensing of air travel organisers.
- Air safety, both airworthiness and operational safety, including the licensing of flight crew, aircraft engineers and aerodromes, and the certification of UK airlines and aircraft.
- Consumer interests, private aviation requirements, economic and scientific research, the collection and publication of statistical and economic data, and the consultancy and training for overseas administrations.
- The ownership and operation of eight aerodromes in the Scottish Highlands and Islands.

Certain other civil aviation responsibilities, including international relations, the investigation of aircraft accidents, policy on aircraft noise and national airports development, and the co-ordination of aviation security, have been retained by the Government through the Department of Transport. On civil aviation matters generally the CAA acts as the Government's expert adviser.

The CAA currently employs 7,000 staff, 78.5% of whom are in the National Air Traffic Services, 9.5% engaged in safety and 1.5% on economic regulation. The Authority's responsibilities require the staff to be highly qualified in their particular fields, including engineers, air traffic controllers, scientists, pilots, economists, doctors and lawyers. Only 20% of the staff work in central London, the remaining 80% working at various locations scattered throughout the United Kingdom.

**National Air Traffic Services**
The National Air Traffic Services (NATS) was established in 1962 as a joint civil/military partnership. Its function is to secure the safe, orderly and speedy flow of air traffic within the United Kingdom airspace, and to provide air traffic services for aircraft flying over the northeastern quarter of the North Atlantic, a responsibility assigned to the UK by the International Civil Aviation Organisation. NATS also provides air traffic services for all British Airports Authority airports, at some local authority airports (Cardiff, Birmingham, Manchester and Belfast) and at the CAA's own aerodromes in Scotland.

The whole range of air traffic control and advisory and information services in UK airspace

*Left and Below:*
**The new visual control tower at Gatwick Airport, opened in 1984. The tower is 45m high and is the tallest in the country.** *CAA*

*Right:*
**Directflight directors and chief pilot with their Partenavia P.68B Victor, G-BGEM, in front of the control tower at Norwich Airport.** *Directflight*

*Far right:*
**An en-route sector suite in the operations room of the London Air Traffic Control Centre at West Drayton.** *CAA*

*Below right:*
**Heathrow Airport radar tower.** *CAA*

TRANS WORLD

is provided from three air traffic control centres. The London Air Traffic Control Centre (LATCC) at West Drayton, with a sub-centre at Manchester, looks after traffic over England and Wales up to the Scottish border (55°N). At Prestwick there are two centres — the Scottish Air Traffic Control Centre (ScATCC) dealing with movements over Scotland, and the Oceanic Area Control Centre (OACC) which looks after trans-Atlantic air traffic in collaboration with Canadian and United States centres.

There are about 100 radar displays at the London, Manchester and Scottish centres which furnish basic information on air traffic, and a network of some 100 navigation aid stations provides radio signals which aircraft use for en-route flying, determining position and locating aerodromes.

Four types of aid are currently used: non-directional beacons (NDB) — en route and at aerodromes; VHF omni-directional ranges (VOR and DVOR); Distance Measuring Equipment (DME); and Instrument Landing Systems (ILS). ATC communications services are provided by air-ground and ground-ground systems, with the ground-ground national and international telephone network centres at LATCC at West Drayton, and telegraphic communications controlled by the Civil Aviation Communications Centre located at Heathrow. Responsibility for the government's aviation obligations in the international sphere is vested in the Department of Transport for civil matters and, on its behalf, NATS discharges the operational, technical and financial ATS commitments arising from UK membership of the International Civil Aviation Organisation (ICAO), the European Organisation for the Safety of Air Navigation (Eurocontrol) and the NATO Committee for European Airspace Co-ordination (CEAC).

The CAA, through NATS, also offers three important information services. The Aeronautical Information Publication contains information on almost everything a pilot may need to know about flying from, to, or over the UK. The UK Aeronautical Information Service (AIS), based in Middlesex, publishes Notices to Airmen (NOTAM) to give up-to-date information to pilots about anything which might affect flying, and air navigation maps and charts of the UK are available to the standard laid down by ICAO.

### Controlled Airspace

With an area of almost 1,000,000sq km, NATS provides Controlled Airspace in the form of a legally protected structure for the main streams of airline traffic. To achieve flexibility for flying outside this space, it has developed a number of subdivisions with different rules and requirements.

Controlled Airspace includes the Terminal Areas (TMAs) which enclose the main airports and groups of airports. It contains the airways which carry the principal flows of commercial, passenger-carrying traffic within the UK. The airways are corridors usually about 16km wide which extend from about 1,500m to 7,500m above sea level. Within a TMA, comprehensive air traffic control measures are used to ensure that arriving or departing traffic can operate safely into and out of the Control Zones (CTR) surrounding the airports below.

Between 7,500m and 20,000m of UK controlled airspace, civil traffic is subject to a full and mandatory ATC service. Below 7,500m Special Rules Airspace provides for a variety of situations where NATS offers an advisory service.

In the northern sector of the North Sea, including the Shetlands, a specialised service is available to the large force of helicopters engaged in the support of off-shore oil exploration and production, for which NATS uses the facilities at Aberdeen and Sumburgh Airports, as well as Ministry of Defence radars at Buchan in Scotland and Saxa Vord in the Shetlands.

### Economic Regulations

As required by the Civil Aviation Act 1982 the CAA's economic objectives are to ensure that British airlines satisfy public demand at the lowest possible charges consistent with a high safety standard, reasonable financial returns and the sound development of British air transport. Before a British airline can embark on a scheduled or charter service, it must have an air transport licence issued by the CAA. Such a licence is awarded only after the airline has satisfied the Authority about its financial and operational fitness. If another airline or other eligible third party objects to a licence application (which almost invariably is the case) the CAA has to hold a public hearing and publish its decision.

In 1973 the CAA began to licence air travel organisers. An Air Travel Organisers' Licence (ATOL) is issued and is designed in part to protect the public from possible loss and, in particular, against the risk of being stranded abroad through the financial failure of air travel organisers. An important function of the Economic Regulation Group is to advise and assist the Department of Trade, which has prime responsibility for negotiations with other countries in respect of air services.

*Above:*
**A general view of the operations room at the Scottish Air Traffic Control Centre, Prestwick. The operational units are arranged in four rows with the outer rows consisting of control suites, each containing a planning and radar position. The units in the centre of the room provide back-up services, including the Flight Data Processing.** *CAA*

*Left:*
**Controllers at work in the Oceanic Area Control Centre at Prestwick, which is responsible for providing air traffic services to aircraft flying in the eastern part of the North Atlantic.** *CAA*

*Above:*
**The terminal building for oil-related traffic at Sumburgh Airport in the Shetlands, developed by the CAA at a cost of £32million.** *CAA*

*Below:*
**Training of Controllers at the CAA College of Air Traffic Control at Hurn Airport.** *CAA*

## AIRWAYS IN UK AIRSPACE

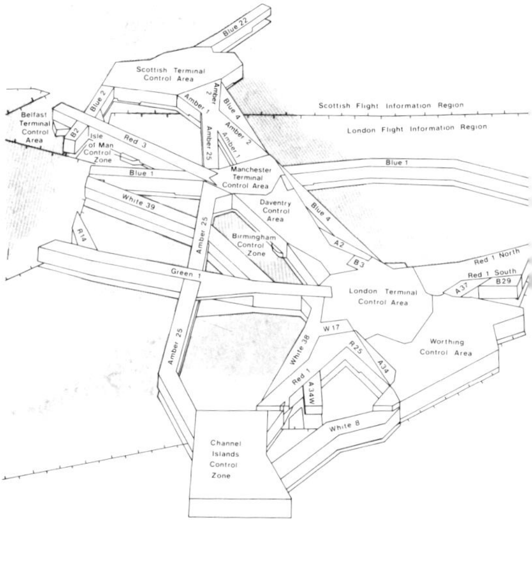

## :AA Scottish Aerodromes

he CAA manages eight aerodromes in Scotland – Sumburgh and Inverness Airports, and the erodromes of Benbecula, Islay, Kirkwell, Stor-oway, Tiree and Wick. Due to their remoteness om large centres of population and only a nodest amount of normal domestic traffic these erodromes as a whole do not generate sufficient evenue and are likely to continue to operate at a oss. Their justification lies in their social nportance to local residents for providing a vital nk with more populous centres for business and velfare, including the flying ambulance service vhich operates 24 hours a day, seven days a veek. In recognition of this community aspect, ne Scottish Office provides a grant to cover the uthority's losses on its operating costs.

The authority's airport in the Shetlands, umburgh, was extensively developed by the AA to meet the sudden North Sea oil boom in ne 1970s, and now plays a significant role in ıpporting oil industry traffic, not only as an irport but also as an air traffic service unit andling air traffic around the Shetlands and ver the North Sea. There are signs however that ne advent of the long range helicopter is eginning to diminish Sumburgh's importance as centre for oil-rig traffic.

## ccent on Safety

he UK has one of the highest civil aviation afety levels in the world and constantly strives or improvement, helping to ensure that the ame high minimum standards can be set by ther countries whose aircraft fly into, out of or ver the United Kingdom.

The CAA's Safety Services include an perational Division, which deals with aircraft perations, and aerodrome and flightcrew censing; a Medical Branch; and an Air-orthiness Division, which deals with the aircraft nd ground engineering.

Before anyone can set up in business as an irline or air taxi operator he must first obtain an ir Operator's Certificate (AOC) or, in the case of gricultural aviation companies, an Aerial Appli-ation Certificate. The granting of an AOC onfirms that the operator has the necessary ıanagerial and technical resources to ensure afe operations.

The CAA sets the training, experience and esting standards for all UK flightcrew licences, nd the Medical Division is concerned with the hysical and psychological fitness of aircrew and ir traffic controllers. The CAA licenses those UK ivil aerodromes which are used for public ansport of passengers or for flying instruction. uch aerodromes must meet prescribed safety andards as regards physical characteristics, aerodrome marking and lighting, and fire and rescue services.

The Airworthiness Division employs about 370 staff, 200 of whom are surveyors — roughly half of them dealing with the design of aircraft, and the other half dealing with the construction, servicing and maintenance of aircraft. About 50 surveyors are stationed permanently at 13 out-stations around the country, and another 15 surveyors are based at nine offices overseas from which the Authority provides advisory services on airworthiness matters to a number of foreign countries on a repayment basis.

Broadly speaking no civil aircraft on the British register can fly unless it holds either a valid Certificate of Airworthiness (CofA) or a Permit issued by the Authority. Before issuing a CofA, the Authority is required by law to be satisfied that the aircraft is fit to fly. This involves taking into account not only the engineering but also the flying characteristics of the aircraft. These objectives are met first by setting the standards in British Civil Airworthiness Requirements. These are published either as BCAR Sections or (as is the case for large transport aircraft, sailplanes and powered sailplanes, engines, propellers and auxiliary power units) by adopting the Joint Airworthiness Requirements (JAR). The latter requirements have been adopted as an inter-national standard by the leading aviation authorities in Europe. With the co-operation of British airlines the CAA has set up the Civil Aviation Airworthiness Data Recording Pro-gramme (CAADRP), which constantly monitors performance and operational standards.

The CAA is obliged under the Civil Aviation Act to consult the Airworthiness Requirements Board (ARB) on significant airworthiness matters. The ARB is an independent statutory body, whose members represent a cross-section of the aviation world and are nominated by representa-tives of manufacturers, airlines, insurers and pilots.

The CAA Flying Unit, utilising a fleet of two BAe 748s and a BAe 125, is based at Stansted Airport and is responsible for all flight inspection work and techniques necessary for the cali-bration of en-route and airport navigation aids, and for investigations into the performance of operational and new navigation aid, radar and communication systems.

The Civil Aviation Authority spent about £3.6million in 1983/84 on its research and development programme, most of which was directed to improving safety and the air traffic services. The total national R&D programme in support of civil aviation operations, which involved an expenditure of about £15million in 1983/84, is reviewed regularly by the Civil

Aviation R&D Programme Board, comprising representatives of airlines, the aircraft and avionics industries, the Department of Industry, MoD(PE) and the British Airports Authority, as well as the CAA.

---

**Civil Aviation Authority**
CAA House,
45 Kingsway,
London, WC2.

***Below:***
**Firemen at the CAA Fire Service Training School at Tees-side Airport learning how to deal with a burning aircraft.** *CAA*

***Centre and Bottom***
**The CAA utilises one BAe 125, appropriately registered G-CCAA, and two BAe 748s on flight inspection and calibration work from Stansted Airport. The 748s are registered G-AVXI and G-AVXJ.** *CAA*